On Capitalism and Inequality

Robert U. Ayres

On Capitalism and Inequality

Progress and Poverty Revisited

Robert U. Ayres
Emeritus Professor of Economics, Political Science,
Technology Management, Novartis Chair Emeritus
INSEAD
Fountainebleau, France

ISBN 978-3-030-39650-3   ISBN 978-3-030-39651-0 (eBook)
https://doi.org/10.1007/978-3-030-39651-0

© Springer Nature Switzerland AG 2020

This work is subject to copyright. All rights are reserved by the Publisher, whether the whole or part of the material is concerned, specifically the rights of translation, reprinting, reuse of illustrations, recitation, broadcasting, reproduction on microfilms or in any other physical way, and transmission or information storage and retrieval, electronic adaptation, computer software, or by similar or dissimilar methodology now known or hereafter developed.

The use of general descriptive names, registered names, trademarks, service marks, etc. in this publication does not imply, even in the absence of a specific statement, that such names are exempt from the relevant protective laws and regulations and therefore free for general use.

The publisher, the authors, and the editors are safe to assume that the advice and information in this book are believed to be true and accurate at the date of publication. Neither the publisher nor the authors or the editors give a warranty, expressed or implied, with respect to the material contained herein or for any errors or omissions that may have been made. The publisher remains neutral with regard to jurisdictional claims in published maps and institutional affiliations.

This Springer imprint is published by the registered company Springer Nature Switzerland AG
The registered company address is: Gewerbestrasse 11, 6330 Cham, Switzerland

Preface

Why did I write this book? The short answer is that Capitalism (with a capital C) is under serious challenge today, because if current economic trends are not stopped and reversed, much of the planet will soon be owned by a small fraction of its population. Our liberal democracy will be replaced by illiberal plutocracy or something much worse. The citizens of our Democracy need to understand why and how the machine that supposedly produces new wealth is actually concentrating existing wealth in the hands of a few ultra-rich plutocrats who already have excessive influence over the political process. People who vote need to understand the motivations and incentives of capitalists on the one hand, and the rules and constraints that drive the capitalist system down this self-destructive path, on the other hand. Finally, all of us need to understand why government needs to create mechanisms for systematic wealth re-distribution, not necessarily by means of taxes, in order to keep the capitalist "game" from running out of players.

This finally brings me to the point. This book is not mainly about the social and economic problems we face. It is mostly about the institutional structures and barriers that support the existing system. Solving problems, in every domain, requires sharing of knowledge, collaboration and cooperation. But capitalism, in its present form, is all about competition, if not outright conflict. It is a game with very few winners and a great many losers. The losers need to be kept in the game, or the system will choke itself to death. Taken to its extreme, one winner will take all. It won't be you.

This book touches here and there on the errors and omissions in textbook economic theory and how they support capitalist practice. It is also partly about conflicting and inconsistent "rights", as between freedom, equality and opportunity vs. the rights of private property and the rights of owners and lenders vs. the rights of renters and borrowers. I hope to bridge some gaps and, more, to provide some guideposts to future leaders. I wish I could do it all in 10 sharply written pages, but I am not that good at simplification. The reality is complex. It needs to be understood as a complex system. Bear in mind that this book will be a lot easier for you to read than it was for me to write. And I hope you have a little fun as you read.

Fountainebleau, France   Robert U. Ayres

Contents

Introduction	1
Before Capitalism	7
The Protestant Reformation and the Rise of "Economic Man"	21
The Age of Enlightenment: 1585–1789	27
Money as Wealth	41
Money as a Medium of Exchange	45
Credit and Banking	55
Money as Gold	65
Money as Printed Paper	73
Capitalism vs. Socialism: A Conflict of Ideas	77
A Conflict of Ideas, Continued	85

The Rise of Corporate Capitalism and Its Champions	101
A Brief History of Financial Booms and Bubbles	113
The Rise and Fall of Bill-Broking and the Central Bank as Lender of Last Resort (LLR)	125
Bubbles and Panics Since 1920	139
Economic Cycles, in Principle	161
Globalization and the Decline of the Labor Movement	169
Mechanisms for Privatizing Profits and Socializing Losses	175
Leverage: How the Rich Keep Getting Richer	191
Is Capitalism Per Se Responsible for Lifting Billions of People Out of Poverty? Or Is It Responsible for Increasing Inequality?	199
Fixing Capitalism: Is It Time for UBI	207
References	223
Index	229

About the Author

Robert U. Ayres is formerly Chair, now retired Professor Emeritus, of economics, environmental management and technology management at INSEAD, a business school with campuses in Fontainebleau, France, Singapore and Abu Dhabi. He is also an Institute Scholar at the International Institute for Applied Systems Analysis (IIASA) in Laxenburg, Austria, and a Kings Professor in Sweden. Professor Ayres holds degrees from the University of Chicago, the University of Maryland, and Kings College of the University of London, with a doctorate in mathematical physics. He has written or coauthored 23 books (before this one), plus more than 200 journal articles, on several academic subjects and co-edited a few others. He is listed in Wikipedia and Google.

Introduction

Somewhere in the dim past philosophers have conceived of the idea that humans are born with "inalienable rights." In the Declaration of Independence of the 13 English colonies in America, these were identified as the right to "life, liberty, and the pursuit of happiness" which is widely interpreted as self-ownership and the right to acquire and protect property. Those aims are vague enough to be generally acceptable and also nearly meaningless when it comes to enforcement. "Liberty" is an especially troublesome idea. If every person has the "right" to do anything he or she wants to do—the state of anarchy—there will inevitably be conflict with other people invoking the same rights. The usual compromise is to modify the notion of "right" to prohibit harming or interfering with the rights of others. This leads to the need for laws, and enforcement mechanisms ("the rule of law") and a minimal government of some sort.

The short list of human rights recognized in the Declaration of Independence was significantly extended by the Bill of Rights (the first 10 amendments to the US Constitution). If you have not read the list recently, the first amendment specifically guarantees freedom of religion, free press, free speech, freedom of assembly, and freedom to petition. The second amendment guarantees the right of citizens to bear arms (in a regulated militia). The third amendment limits the government's right to force citizens to quarter soldiers even in time of war. The fourth amendment limits the government's right to search for and seize evidence or illicit substances (e.g. drugs). The next four amendments pertain to judicial process, the ninth amendment acknowledges that other rights exist and may be important (e.g. the right of privacy in marriage), while the tenth amendment spells out which areas are in the domain of the Federal Government and which are left to the States or the People.

I think it is time to include explicitly another right that we Americans brag about, that we claim distinguishes us from the rest of the world, but that we signally fail to deliver. I refer to the right of *equal opportunity*. Inequality of opportunity has been a major theme of social reformers ever since Thomas More. Some social reformers in the nineteenth century, notably the communists, argued that equal opportunity is inconsistent with private property. Equal opportunity seemed to them to require social (i.e. state) ownership of the "means of production." This focus on state (i.e. government) ownership, vs. private ownership, has become a litmus test for the long-running contest between socialism vs. capitalism as philosophies.

Is that all there is to it? Obviously it is not. The question immediately arises: How does government come to exist in the first place? Does it organize itself? Is it a dissipative structure far from equilibrium as Prigogine might have suggested? What are its limits? Is it a natural consequence of the competitive interactions between persons in different circumstances? How are laws created in different forms of government? Of course, the world, as it exists today, is the not the result of interactions between Hobbesian individuals starting "from scratch." The existing "system" is the consequence of a long and complicated history. Historical changes are certainly attributable to interactions at the individual level, but even more so, to interactions between families, tribes, communities, companies, states, and societies. Our understanding of the interactions between these entities is still quite limited, to put it mildly.

Knowledge is not just a collection of facts. Librarians sort books mainly by category: history, science, economics, politics travel, "self-help," religion, biography, poetry, drama, detective fiction, science fiction, romance. Knowledge is also organized into theories. We have theories to explain observables. Newton's theory of gravitation explained the motions of the stars and planets, but not the motions of electrons and protons. For that, Maxwell's of electromagnetic forces was needed. More complex phenomena require more complex theories to explain them. The most complex phenomena, the behavior of heterogeneous human beings, as both as individuals and as components of various social structures, are still not well understood at all.

The only domain in the social sciences where theory is reasonably well developed concerns the behavior of "actors" (individuals and of business organizations) buying and selling goods and services in exchange markets. Yet those theories are seriously flawed, as this book will point out. One of the crucial flaws is that existing theory focuses too much on competition and the benefits of competition, without giving sufficient weight to the role of cooperation. This overemphasis on competition and its "winners" has led to laws and regulations that promote competition and protect the property of the winners while not sufficiently protecting the losers.

Philosophers have noticed the conflict between "freedom" for individuals, in the anarcho-libertarian sense, and social justice. What political libertarians today mean by "freedom" is minimal government. Some call it the "night watchman" state, where the only role of government is to protect private property from theft. Freedom, in that sense, is fairly close to anarchy. What do the rest of us really mean by "freedom"? The United States of America still calls itself "the land of the free". But that was when land in Oklahoma was given free to new settlers—regardless of the opinion of the native population.

Freedom does not extend to non-payment of taxes, driving on the "wrong" side of the street, smoking cigarettes in public places, shouting "fire" in a theater, or leaving garbage on the street. Some think it allows them to keep their children out of school to work on the farm, or to force schools to teach "intelligent design" instead of evolution. Most conservatives, who want less government and no restrictions on gun ownership, also want to prohibit a variety of private actions, ranging from smoking pot, using contraceptives, abortion, or living and working in this country without "papers." The capitalism we enjoy today allows us to overconsume junk foods (not to mention other kinds of junk) that cause obesity and simultaneously pollute the planet and cause climate warming.

I believe that there is room for compromise. Freedom to choose cannot be absolute, nor can equality of incomes or outcomes be guaranteed. Competition, within limits and subject to rules, has benefits, but cooperation is also essential. Equal opportunity cannot be guaranteed by law, any more than equality of outcomes. The best measure of an economic system is not aggregate wealth or income. The best measure, in my view, is whether opportunity for advancement, for the disadvantaged people in our society, has been increasing or not. By that standard, capitalism American style is a failure. I think the current system can be fixed, and the need is urgent.

Who am I to pontificate on such deep subjects? I am reminded of a story. It may be apocryphal. But it seems someone asked Pablo Picasso how long it took him to create his most famous painting "Guernica," now at the Museum of Modern Art in New York. He thought for a while—a long while—and finally answered his interrogator with a phrase: "My whole life." So, with apologies, here is a capsule history of who I am and how I came to write this book.

From 1954 through 1962, I was studying physics or working as a post doc on research. After the last US atmospheric nuclear test (1962), I was hired at the Hudson Institute to prepare a Report to the Office of Civil Defense on the environmental consequences of a global nuclear war (the report still exists). In the course of that effort, I read the only textbook then existing, on a new subject "ecology," by Professor Eugene Odum at the University of Georgia. Ecology was the buzzword that covered the whole set of interactions between plants, animals, and the physical environment. In 1966, I decided on a career change.

My introduction to economics occurred in 1967–1968, which I spent working at a newly created institution called Resources for the Future, or RFF. RFF was (and still is) a "Think Tank" mostly run by economists, focused on economic issues pertaining to future resource needs and availabilities. I devoted that year to working my way through a new subject, environmental economics, under the tutelage of Allen V. Kneese. Allen became not only my mentor in economic thinking, but my colleague and friend. We wrote several articles and papers together, in that year, on a range of topics involving environmental damages caused by industrial and municipal wastes and air and water pollution.

The paper that mattered most, in terms of later events, was an article published in the American Economic Review (AER) entitled "Production, consumption and externalities: A Materials-balance approach" (1969). What that paper said, briefly, was that all kinds of pollution and waste are direct, unavoidable consequences of the laws of thermodynamics (i.e. physics) and that this scientific fact has strong implications for economic theory that were not—and still are not—reflected in the standard theory as taught in schools and colleges. That paper planted a seed that never quite stopped growing.

But that seed took some time to germinate. After leaving RFF in 1968, I joined another ex-nuclear physicist, Theodore B. Taylor. Together we opened a consultancy in Washington D.C. with the grandiose name International Research and Technology Corporation (IR&T) to undertake studies on various scientific topics for the government. Ted Taylor specialized in nuclear weapons proliferation, safety issues, and nuclear power. I specialized in everything else, but with a main focus on transportation, pollution, and energy issues. Later (with other colleagues) I spent a lot of my time thinking about what later came to be called "industrial metabolism" or "industrial ecology."

For decades, I have worked on the interfaces between science, technology, material flows, and—finally—economic theory again. I think it is fair to say that in the 1990s, I did experience a rather sudden and surprising insight not very different from the epiphany once experience by an Arab named Thomas, (later awarded sainthood) on his journey to Damascus. It was a different insight, of course, but the nature of the experience was similar. What I realized one day (or night) was the simple fact that economic growth is driven by energy, in the sense of doing physical work. But I also noticed that energy (and work) are totally absent from fundamental economic theory. Physicists tend to recognize the truth of this statement instantly. But economists rarely study physics. They see energy and work as "intermediate products" of capital and labor. The fact that energy can neither be created nor destroyed has not yet been accepted.

But this book is not about that, either. I have written about it elsewhere. It is about why financial capitalism is failing to deliver the American Dream, and what can be done about that.

Before Capitalism

Around 12,000 years ago, the glaciers were melting, and forests were growing on the newly exposed hills. Game was plentiful, and our ancestral hunter-gatherers were self-sufficient and reasonably well-fed. But as the glaciers retreated, the climate became warmer and drier, and the lowland forests were replaced by scrub and grasslands, forcing the inhabitants to become either peripatetic animal-herding nomads or to build defensible settlements and grow crops that require some time to germinate grow and ripen. From time to time, the nomads attacked the settlements. That cultural conflict still has echoes in the world today.

But by 8000 B.C., the *H. sapiens* population had outgrown the food supply available for hunting and gathering from forests, in several regions. Where both game and grass were scarce, our human ancestors learned to plant seeds and wait for them to grow. They become farmers. This happened first and primarily in the Mediterranean basin and specifically in the so-called fertile crescent where the Tigris and Euphrates flow into the Persian Gulf. This happened first in that area because it or nearby areas were the sources of the wild progenitors of most of the major food crops of today (Diamond 1998). Farming spread, gradually, from that area to other major river valleys such as the Nile, the Indus, the Brahmaputra, the Mekong, and the Yangtze-Kiang. Those river valley lands were fertile thanks to the annual floods, and there were no deep-rooted plants (perennial grasses and scrub) to make cultivation difficult.

There is a story, now largely discredited, that farming became established because it was easier and more productive than hunting. The truth is otherwise. We now know this from the smaller sizes of the skeletons of agricultural workers as compared to their hunter-gatherer forbears. However, it is true that once farming got started in fertile areas, such as river deltas that receive annual silt deposits from spring floods, food production did grow rapidly. Storage of surpluses also increased.

Agriculture spread to dry grasslands much later. This required the invention of plows, yokes, and harnesses to enable tamed animals, like bullocks, to do the plowing. Farmers—or their rulers—learned to store grain (seed) crops from year to year. Later still they learned to store water and irrigate during dry seasons. Growing population pressures, and natural disasters, also led to tribal organization, tribal boundary disputes, walled cities, and—when the local food supply ran out—wars of conquest. The human population grew faster, in many years, than the agricultural food supply.

Urban civilization, as we understand it, probably began around 4000 BCE in the ancient Mesopotamian land of Ur, located near the mouth of the Euphrates River, near it is confluence with the Tigris River in present-day Iraq. (There are no earlier remains of cities in Mesopotamia, probably because they were destroyed by floods.) Their language, now identified as proto-Indo-European, was the common ancestor of all modern European languages, as well as Urdu and Punjabi. Sumer had organized agriculture and a written language (cuneiform). It was taught in scribe-schools by 3500 BCE (earliest estimate). Icons or symbols—today we would call them letters of the alphabet—were first created by pointed sticks in wet clay. In fact, the written language may have already existed in that part of the world for hundreds (if not thousands) of years before the oldest surviving examples.

But in 3100 BCE (earliest estimate) the first system of abstract symbols for numbers was also created, and simultaneously, or soon after that, a system of bookkeeping for quantities of commodities (e.g., bread, beer, oil, dates, sesame, and wheat) followed. Each commodity had its own symbol, represented by a small clay token with a specific shape—rather like the letters of the alphabet or the icons used on personal computer keyboards today. At first, quantities were recorded in the form of physical piles of tokens. Later, records were kept by impressing the appropriate token on a moist clay tablet next to a number symbol. Our alphabet originated in Sumer, as did mathematics, the art of building, the wheel, and the hours of the day.

The earliest written "story" that we know of, in human history, is the *Epic of Gilgamesh*. The oldest Sumerian version dates from 2150 to 2000 BCE. The later, Akkadian version dates from 1300 to 1000 BCE. It was found in the form of 11 clay tablets in a library in Ninevah, an ancient city in Mesopotamia (now Iraq). The story describes the efforts of Gilgamesh, the legendary ruler of Uruk (who is 2/3 god and 1/3 human) to build a perfect wall around his city (Fig. 1).

Before Capitalism 9

Fig. 1 Foundations of a Neolithic dwelling at Tell es Sultan, in Jericho. Wikipedia.org, photo by A. Sobkowski

The wall—probably similar to the wall of Jericho—was supposed to separate civilization (inside the wall) from wild ("sauvage" in French) nature outside. Indeed, when the wall was finished, the city would become the ruler and manager of its neighboring farms and villages. Gilgamesh was obsessed with increasing the productivity of the workers on the wall. The workers on the wall become virtual slaves as described in Tablet 1 (Sedlacek 2011, p. 21).

The young men of Uruk he harries without warrant
Gilgamesh lets no son go free to his father
Gilgamesh lets no girl go free to her bridegroom
The warrior's daughter, the young man's bride

The actions of Gilgamesh in this part of the story strongly resemble the actions of present-day corporate raiders or leveraged buyouts (LBO's) that have taken over a struggling firm and who engage in "financial engineering" and "restructuring" to make the balance sheet look more attractive to investors. The villagers (peons) outside the wall would no longer be familiars and equals, but subjects of the ruler—needed to do the work—but otherwise kept at a distance. They were relegated to the bottom layer of a hierarchy managed by military and civil officers, viziers, clerks, and tax gatherers reporting to the king (Mumford 1961).

Even Greece, where we suppose "democracy" was invented, it seems that their greatest philosopher (Plato) had a different vision. A recent critic points out the following:

> In his vision of an ideal state, Plato does not allow guardian families to raise their own children. Instead they hand them over to a specialized institution immediately after birth. This is similar to the dystopias in Aldous Huxley's "Brave New World" and George Orwell's "1984". In both novels, human relations and feelings (or any expression of personality) are forbidden and strictly punished.... (Sedlacek 2011, p. 22)

There is a lot more in the epic of Gilgamesh.

Later, in the Hebrew *Old Testament*, the "evil" farmer Cain (who grows crops) murders his "good" brother, the shepherd Abel. Cities in the Old Testament were regarded (by the nomadic Jewish prophets) as evil, corrupt places of sin and decadence, such as Sodom and Gomorrah.

Fifteen hundred later (2100 BCE), the city of Uruk in Mesopotamia had a population of over 60,000 on a land area of 5.5 square km, on the banks of the Euphrates River. It was governed by a semi-divine king, who controlled the army, and a chief priest who controlled the bureaucracy of civil servants (clerks, scribes, and accountants) in the Temple, which was located in the ziggurat in the center of the city, replete with mosaics and magnificent gardens. On the top was a temple to the city's deity Nanna, Goddess of the moon (Fig. 2).

Most of the fields, herds of animals, and marshes (fish farms) were owned by the Temple, which was dedicated to the God Nanna and his spouse Ningal (Martin 2014, pp. 38–44). The citizens worked and must have received rations—proto money—for their work because coins had not yet been invented.

But circa 2000 BCE, there was a major invasion of the territory of Ur. Ur was sacked, along with 16 other cities in the territory of Sumer. The invasion and destruction of Ur had an effect on the evolution of religion, at least polytheistic religions with many personal or local gods. Ur declined thereafter. The tribal leader known today as *Abraham* ("father of the Jewish people") left Ur, with his retinue, going north (upriver), around 1800 BCE. His people, the Hebrews, were the first monotheists. All three of the modern major religions, Jews, Christians, and Muslims, are known as *Abrahamist*, given their strong historical links to the City of Jerusalem and the so-called Holy Land.

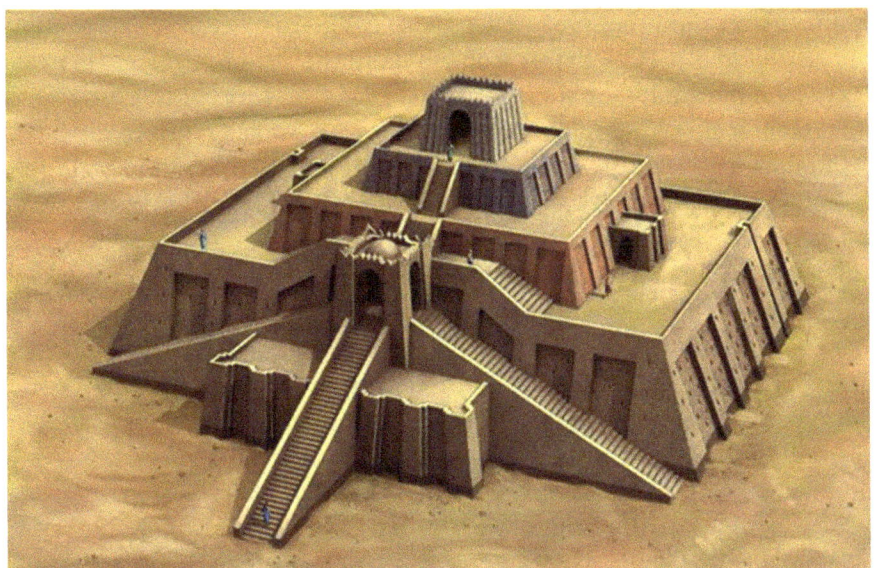

Fig. 2 The Great Ziggurat of Ur (Google Earth)

The Origins of Ethics and Morality

The most familiar, if not the first, code of ethics was the "Ten Commandments" of the Hebrews. It was allegedly delivered to their leader, Moses, by God, on the top of Mount Sinai. In those days, women were the property of their husbands. (In the Koran a woman is (still) valued at only 2/3 the value of a man.) Gender equality is a relatively recent idea, and only in the more economically developed countries.

The most "advanced" countries in the world today are struggling to rectify an age-old imbalance between the rights and privileges of men *vis-a-vis* the rights of women. Women only got the right to vote in most countries in the twentieth century. The Scandinavian countries are the most advanced in terms of legal rights (e.g. re marriage and child support) and equality of income. But there are still major parts of the world where females are not treated as legal equals of males. As noted above, orthodox (Wahabi) Islam—Saudi Arabia, in particular—is the most obvious retrograde example.

The prehistoric origins of gender imbalance are fairly obvious. They stem from the biological fact that women in prehistoric times had to spend most of their adult lives—usually starting soon after puberty—being pregnant and/or caring for babies or small children. This is a direct consequence of the fact that hominin children (unlike the offspring of four-legged animals) cannot care for themselves until they are several years old, and that "growing up" time for human children is 10–12 years (roughly until the age of puberty). The caregivers were the adult—or adolescent—females. The females, in turn, needed protection while taking care of infants.

In nomadic tribes, the males had to defend the group and to obtain food by hunting for game. However, in forested areas some of the "gathering"—collecting nuts, fruits and edible roots or fungi—could also be done by women and small children, as long as it could be done near the cave or stockade. Hunting and defense required weapons, and the use of weapons required training and practice. This resulted in military-style organization, led by men. So it was natural that prestige and leadership in hunting societies became associated partly with age and experience and partly with physical prowess and skill with weapons. There is some argument about the gender roles in hunting and gathering societies, but mythological hunters were male.

Here it is important to say a few more words about the evolution of ethics and morality, i.e. the behavior of people (men and women) towards each other. The "Ten Commandments" reflected a society of authoritarian patriarchal families concerned with their own welfare, but required to avoid certain behaviors that would create stresses in the larger society. The existence of absolute rulers was taken for granted. Gilgamesh was an absolute ruler. Rulers were abjured by Prophets to be less oppressive, especially with regard to the treatment of widows and orphans. A rich man "of many shekels" should not prey on a poor man of one shekel.

The guidance was strictly negative: do not do this, do not do that. What was missing was positive guidance as to relationships. There are plenty of snippets of behavioral advice from prophets and wise men, from Buddha to Zarathustra, but they focused more on human relationships with God (or Gods) than on human relationships with each other. Buddha, for example, focused on individual enlightenment in isolation, and the practice of yoga. The Mosaic law specified limits for response to an injury "an eye for an eye and a tooth for a tooth."

Rules

The core of ethics today is the rule most memorably set forth by Jesus Christ of Nazareth: that a man should care for God with all his heart and mind and care for his neighbor as he would care for himself. Jesus exhorts his followers to "Do unto others as you would have others do unto you" or words to that effect in Hebrew. This advice was not new when Jesus preached. From the

Analects of Confucius, a similar flicker of light emerged, even earlier: Tsu-kung asked the master, "Is there a single saying that one can act upon until the end of one's life?" The Master answered with a question: "Would it be reciprocity? What you do not wish done to yourself, do not do to others." The most advanced philosophers went beyond mere reciprocity. Jesus Christ added, "Turn the other cheek also." Even earlier a nameless sage wrote as follows: "Recompense with good the man who wrongs you." Another Chinese sage, Mo Ti, said that T'ien (Heaven) loved men and that all men should love one another … a doctrine of universal care (love) is the basis of ethics.

Morality is central both to religion and to the problem of debt. Almost every person alive thinks that the repayment of debt is a moral duty. That idea seems to have an ancient origin as David Graeber has explained at great length (and depth) in his magnum opus (Graeber 2011). One quasi-religious explanation of debt is that it started with the first recognition of continuity of society: we who live now are alive thanks to the life of our ancestors, hence we are indebted to them. That kind of debt is usually thought of as a duty. Performance of the duty is sometimes called "redemption," another word for pay-back. Christ was the "redeemer" in the sense that he relieved us of our debt to our ancestors.

Over the past several 1000 years, debt has gradually changed from religious obligation to our ancestors. More and more, it is associated with money to be paid. Yet the argument in modern finance theory is that the possibility of a default is just a quantifiable risk, not a moral duty, and that the magnitude of the risk of default is simply reflected in the rate of interest charged. Lenders should set the rate of return on every debt in relation to its risk. This is a huge change from historical views on the subject.

Mahayana Buddhism (first century AD) introduced the notion of Bodhisattva ("being of enlightenment") who sacrifices his personal salvation to help other creatures by acts of love and compassion. The stoic philosophers, such as the slave Epictetus, advocated religious resignation, forbearance, and love toward all men. St. Francis of Assisi, a 1000 years later, also devoted himself to a life of poverty and service to the poor, the sick and the lepers. He founded the Franciscan order. In more recent times, we see many more such examples of altruism such as Mother Teresa.

Notwithstanding the above, some economic theorists (such as Gary Becker from the Chicago School) still question whether altruism is a factor in economic relationships. The question arises because so much of economic theory is predicated on the assumption that markets reflect the actions of rational persons whose calculations of utility are based solely on self-interest (greed). In short, neoclassical economists tend to assume that greed is the fundamental and only driver of all human activities.

The Sumer equivalent of Noah, the Jewish ark-builder who saved all the living creatures from the great flood, was named Utanapishti. In his words, quoted in the Epic of Gilgamesh:

All the silver I owned I loaded aboard,
All the gold I owned I loaded aboard
All the living creatures I had I loaded aboard
I sent on board all my kith and kin
The beasts of the field, the creatures of the wild,
and members of every skill and craft. (Sedlacek 2011, p. 36)

Unlike the Hebrew ark-builder, Noah, Utanapishti presumably used gold and silver (money) for exchange purposes. So the Gilgamesh epic presupposes the existence of money and markets. Yet during the lifetime of Jesus (according to the Bible), merchants and money changers were perceived by Jesus and his followers as despicable, hence it was necessary and admirable to expel them from the Temple in Jerusalem (Fig. 3).

Fig. 3 Jesus cleansing the Temple (El Greco), National Gallery of Art, Washington D.C. (online collection)

Asceticism, poverty, and self-abnegation are greatly—but not universally—admired. One of the most admired of all was Saint Francis of Assisi. Celibacy and chastity (except for procreation) have been important in Christianity, classical Hinduism, and Buddhism, but not in Judaism or Islam. Most cultures have had trouble reconciling money with virtue. The conflict is visible everywhere today.

The Class System

The historical division between the "nobles" and the "commons" probably began with the distinction between urbanized, landowning settlers and the roaming nomads. Whatever its origin, it is still alive and well in the structure of most Western republics. There is an upper house (the House of Lords, the Senate) and a lower house (the House of Commons and the House of Representatives). In both cases, the lower house initiates and the upper house pontificates (and may veto). The same division is to be found in the military services. There are two classes: officers—supposedly "gentlemen"—who give the orders (passed down the "chain of command" from above) and the lower ranks at the base of the hierarchy who obey the orders (or else). Officers do not socialize with ordinary soldiers.

In universities, today, there is a comparable divide between teaching and research faculty (professors) who are mostly garnished with a PhD degree, *vis-a-vis* the lower ranks, such as students, teaching assistants, librarians, janitors, cooks, and so on. Faculty members decide what to teach and how to teach it. The rest take orders. Similarly, in most corporations or partnerships there is a gulf between the owners—founding shareholders or partners, who invested money (or who inherited shares from a parent)—*vis-a-vis* the employees and wage earners. The shareholders or partners receive profits (dividends) and possibly bonuses in addition to salaries. On the other hand, the employees receive wages when the company prospers, but may be let go, with nothing to show for their efforts if the company runs into difficulty. Of course, the reality is more complex than the above characterization, but the asymmetry of "rights" between the upper and lower classes is recognizable.

The same asymmetry of rights holds in the world of finance. The prime lenders (banks) have legal rights superior to the rights of borrowers, while the senior bondholders have rights superior to subordinated bondholders or (at the bottom) the equity holders. When there is a financial problem at a company, the government collects taxes owed, in full, before the next claimants (employees) get paid, then come the bondholders, in order of seniority. All the bondholders get paid before the equity holders. Even the shareholders get paid before the taxpayers, who invested in the infrastructure that enabled the company to do business. This maldistribution of assets and claims bears no relation to the causes of the default.

This asymmetry of responsibility between lenders and borrowers is a leftover from the feudal asymmetry of rights between lords and commons. As will be seen later, it is a primary reason for the pattern of "bubbles" and "busts" that characterize economic history. I will come back to this topic again.

Social Structure and Land

We do not know in detail how neighboring tribes in the distant past managed to come together to become regional kingdoms, although it seems clear that the combination of physical and tactical leadership skills was involved, together with an instinct for cooperation, as well as competition among villages and local warlords. Must have been at the core. Leadership, in war or in village life, also requires trust. What we know of the early history of Egypt, Crete, Mesopotamia, Persia, India, or China, before written records, is what can be inferred from the realm of legend. Conquest was usually the first step. Obviously an integrated country with a central ruler is likely to be stronger and more durable than less organized neighboring countries. This presumably accounts for the rise of empires. But successful government in the long term, within the empire after conquest, was (and is) dependent on power-sharing. Power-sharing requires negotiation and cooperation, both of which require trust.

The origin and development of social organization from tribe to empire is still far from clear. Some tribes (like the early Greeks?) treated all boys as "citizens" of equal rank. Most tribes have a hierarchical organization, with a definite leader or "elder." In some tribes men and women have equal voices, in others not. Greek legends also mention Amazons, warlike women. But the method of leadership selection and succession varies quite a lot. Some tribes have elections (the origin of democracy?) but many do not. Most tribes that we know of are led by men, presumably due to tradition, based on superior physical prowess in hunting or fighting, as noted.

Notwithstanding the sad and continuing record of human inhumanity, there is increasing evidence that voluntary cooperation between diverse ethnic or religious groups almost certainly played an important role in social evolution (Boehm 1999; Wilson 1966; Wilson 1975; Wilson & Wilson 2007). The examples of beneficial outcomes of interactions between different ethnic and religious groups are less well-known, for obvious reasons. The existence—and success—of the United States of America, and of Canada and Australia, as multi-ethnic and multi-religious nations, notwithstanding continuing racial problems, makes the point.

We also know that, in the feudal past of Europe and Russia, between the end of the Roman empire and the rise of nation-states starting in Europe (c. 1300)—Asia much earlier—power was tied to land ownership. Each of the social classes held land "en feu" (in fee) to the next higher rank, from peons to Knights, Barons, Earls, and Dukes. They owed their livelihoods to the level above (and provided military manpower, if needed) while the upper levels of society, living behind castle walls, offered protection and livelihood to the lower. But the rise of walled towns ("burghs") created—or enabled—a new class of merchants (burghers, bourgeois) to live and prosper behind town walls but outside the feudal system. The feudal system started to crack because of that change, amplified by the bubonic plague that arrived two generations later (c. 1340), plus the spread of ideas generated in the new universities.

Around 1300 A.D., ambitious kings, like Philip IV "Le Bel" of France and Edward I of England started claiming "divine right" to rule with absolute power and without sanction from the barons. As the feudal system began to crumble, land-ownership became legalized, leaving a large part of the population landless. Landless people before the industrial revolution existed either as "peons" or "serfs" working the land for the "noble" landowner, or making a precarious living outside the system, grazing animals on "common land" or poaching in the noble's forests, like the legendary "Robin Hood." Landless men also found employment as soldiers, working for the nobles. The enclosure movement based on wool production and exports, for money, ended the free peasantry of England, with traditional rights of occupation and use of common land.

Many villages and some towns existed only to serve the medieval magnates. But others had more earthly functions such as trade and manufacturing. The walled towns in Europe, and doubtless also in India and China, were not only places where displaced peasants settled if allowed (or not) but also places for artisans and merchants to congregate for protection, markets, and especially for synergy.

The synergy is critical. There is a strong tendency for people with specialized knowledge, and skills, to be preserved or concentrated near each other, usually in town centers. Again, the instinct for cooperation is at work. Gilds are for cooperation among craftsmen to maintain stability. Towns, in turn, located themselves trading hubs. This tendency led to regional specialties, e.g., Venice for glass-making, France for wine and perfume (and artillery), Holland for tulips, windmills, and hydraulic engineering, Antwerp for metal-working and diamond cutting, the Ruhr valley and Pittsburgh for steel, Dresden for "China" ware, Switzerland for clocks, Akron for tires, Detroit for cars, and "Silicon Valley" for electronics and software.

At first the European craftsmen formed themselves into guilds, for the purpose of regulating and monopolizing trade. Entry to a guild started with apprenticeship, working under a master craftsman. The apprentice became a journeyman guild member after he (no women allowed) could produce a "masterpiece," i.e., a piece worthy of a master craftsman. A big town may have had 100 guilds. Later the traders formed merchant guilds that eventually absorbed the craft guilds. Residence (citizenship) in the town was a protected right; citizens in the autonomous towns were exempt from many of the laws of the territorial rulers. They had their own courts, for instance.

The social divide between descendants of absentee landowners (who made land ownership a condition of voting) and landless laborers who later became factory workers (and only got the vote much later) still exists. The peons of Eastern Europe were still legally tied to the land until the Russian Revolution and the end of World War I (Tawney, 1931 [1952]). In virtually all armies, there is a social divide between officers (supposedly "gentlemen") who give the orders and ordinary soldiers, and non-coms who are supposed to obey the orders and never talk back. The two classes of soldiers are still treated very differently.

The monarch at the top of a hierarchy exerts supreme authority over all of their subordinates. But the authority is usually exerted through a chain of intermediaries. In a large hierarchy, such as an army or a big corporation, there may be a dozen or more intermediate levels, each one exerting authority over the one beneath while being subject to commands from the level above. (In the army this feature is called "chain of command.")

The first successful non-violent check on royal power in Europe was the Great Charter (Magna Carta) of 1215 CE, between King John "Lackland" and his barons. It protected Church rights and protected the male population (not only the nobles) from "illegal" imprisonment. It said that "No freeman shall be arrested or imprisoned … except by the lawful judgment of his equals and according to the law of the land." That was the origin of the doctrine of presumption of innocence. The Magna Cart also put limits on the monarch's right to tax his vassals. Even though these first "rights" were soon annulled, it may have been the first time when "rights" of any kind were established in law.

In actual fact, the Great Charter is more influential in retrospect than it was at the time. It did not protect the English peasantry, nor did it protect native peoples in conquered lands, slaves, or women. Under British common law, women were virtually the property of their fathers or husbands until the twentieth century or so. As noted already, women citizens only got the vote in the twentieth century, after a long high-profile campaign. They still do not get equal pay for equal work.

The seeds of new knowledge-based power centers emerged in the twelfth century. The oldest universities in Europe are Bologna (1088), Oxford (1096), Paris (1096), Montpelier (1150), Cambridge (1209), Salamanca (1218), and Padua (1222). In the thirteenth, fourteenth, and fifteenth centuries, many more universities were founded in Europe. Even the monasteries opened up somewhat. University-based scholars—mostly associated with monastic orders, to be sure—began to export their knowledge to the world, in books. Pope Clement IV supported their efforts to collect the learning of his time. Friar Thomas Aquinas (1225–1274) was the most famous of the scholastic philosophers of the period; he was the prime interpreter (and promoter) of Socrates.

Questions about astronomy were among those asked for the first time. In 1543, Nicolas Copernicus published a monograph "De Revolutionibus Orbium Coelestium" (On the revolutions of the celestial spheres) which argued that the sun, not the Earth, was the center of the universe. The Church authorities grumbled, but God did not strike down that questioner. Copernicus was followed by Tycho Brahe and Johannes Kepler (whose two laws of motion were published in 1609), and a century later, the Copernican theory itself was struck down, not by the Church, but by Isaac Newton. Copernicus is shown below stargazing (Fig. 4).

The change, as characterized by Harari, was a deep shift from believing that everything worth knowing was already known, to a new and different belief that there was new knowledge of importance to be discovered (op cit). The discoverers of new islands and new harbors on the coast of Africa were folk heroes of the age. They also helped to dislodge the idea that there was nothing new to be seen or learned under the sun. And more discoveries followed the searchers.

A litany of dates associated with printing and book production does not fully reflect the historical importance of the printing press. The spread of books was accompanied by increasing urbanization, increasing literacy, more efficient government and more efficient financial institutions. Beginning in the sixteenth century, there was a gradual, but significant, increase in incomes and capital accumulation, in the countries of northwest Europe (van Zanden 2009). This combination of trends concentrated in northwest Europe helped to prepare the way for the industrial revolution that came two centuries later.

Fig. 4 Astronomer Copernicus or conversations with God by Matejko (Wikimedia commons)

The Gutenberg Bibles and their imitators were of great importance also because, thanks to them and their successors, bibles were soon available in all the vernacular languages of Europe, not only in Latin. This enabled people outside the formal Church hierarchy direct access to (as they thought) the "Word of God." This, in turn, greatly weakened the moral authority of the bishops and clerics *vis-a-vis* the laity as people realized that the use of Latin in the Church was basically a device to keep them ignorant and dependent. This was an important step in the decentralization of power. It effectively paved the way for Martin Luther and Protestantism.

The analogy with mergers and acquisitions in economics is highly suggestive. Most mergers are negotiated, but acquisitions opposed by management do occasionally occur. Most mergers are unsuccessful, due to "cultural differences" according to conventional wisdom. But the successful ones seem to create—or concentrate—more wealth than the unsuccessful ones destroy.

The Protestant Reformation and the Rise of "Economic Man"

The Protestant Reformation was a direct consequence of converging trends. One was the creeping commercialization and corruption of the Roman Catholic Church. Another was the spread of education among towns-people and increasing demand for books in the vernacular languages (i.e. not in Latin). The latter demand was met by Gutenberg and his imitators, resulting in mass production of books (and other printed materials), especially after 1500. The teachings of Martin Luther and Jean Cauvin (John Calvin) were spread initially by preachers, but supported by books. Their teachings changed everything. In fact, there is a strong argument that Protestant doctrines equating wealth with virtue were the incubator of modern capitalism (Tawney 1926).

The Protestant Reformation per se began with the publication of Martin Luther's "Ninety-five Theses", a book that was nailed to a Church door in his town of Wittenberg, Germany. This was an established invitation to debate. In fact, many thousands of copies soon were printed and sold. Without going into detail, the protests were directed at corrupt practices by the Church of Rome, especially the sale of indulgences—"guarantees" of salvation—to lay Church-goers, not to mention the sale of Bishoprics and Cardinal's Hats to wealthy Italians. The real object of the protests was that, during the Renaissance, spiritual services had become items of commerce. (The fact that the money collected from believers was being used—at least partly—to finance the reconstruction of St. Peters' Church in the Vatican is beside the point.)

Once the upper echelon of the Church-of-Rome was widely seen to have morphed into a gang of money-grubbers, the laity needed another moral guide. They found it in money-making. Whereas in the past, religious devotion often involved celibacy, poverty and rejection of worldly affairs, the new form of devotion was to acquire wealth and advertise it. Tawney wrote that Calvinism,

> *"a creed which transformed the acquisition of wealth from a drudgery or a temptation into a moral duty was the milk of lions. It was not that religion was expelled from practical life, but that religion itself gave it a foundation of granite.. The good Christian was … the economic man"* (Tawney 1926) Italics added.

And the wealth should be visible, as a sign that the owner was doing God's work and was assuredly one of God's "elect". In France, Jean Cauvin (John Calvin) was the most effective interpreter of Luther's ideas into practical form. The Puritans in England and the Presbyterians in Scotland spread the new doctrines along with exploration and colony planting.

Max Weber pointed out that, already by Ben Franklin's time, the spiritual content of Calvinism had gone (Weber 1902). It was gone by the middle of the seventeenth century, when the first treatises on economic theory were written by William Petty and others. The old admonition against usury (interest) was gone, too, although the prejudice against Jews remained. Thrift and hard work were already admirable for themselves. Making money from money, was now quite admirable. Benjamin Franklin's contemporary, the moral philosopher Adam Smith, in England, saw it exactly the same way, and wrote a best-selling book "Wealth of Nations" about it (Smith 1776 [2007]).

Before Luther and Calvin, people did not expect to be better off than their parents, nor did they expect that their children would be better off than themselves. One's "place" in society was supposedly fixed by birth. People were taught not to question that reality, because the social structure (designed by God) was not to be questioned. But Luther and Calvin questioned the unquestionable—the Mother Church—and set the ball rolling.

Of course, the old moral paradigm did not disappear overnight. In fact, it still exists. It would not be far off the mark to say that much of the turmoil in the world today is describable in terms of conflict between incompatible moral codes. An interesting characterization of this conflict is due to Jane Jacobs (Jacobs 1992). She characterizes the moral syndrome of urban merchants and traders as "commercial". The other syndrome is applicable to several groups, including the modern equivalents of landless nomads, viz. organized criminals. That might be characterized as the "raider" syndrome.

Jacobs uses the more positive term "guardian." (It might also be called "Machiavellian", in honor of the author of the phrase "*It is better to be feared than loved, if you cannot be both*").[1]

Clearly the commercial/trader syndrome is familiar in the business world, especially exemplified by Ben Franklin and the Pennsylvania Quakers. The "guardian" syndrome is familiar in the armed forces, the police, the religious hierarchy and in team sports. I reproduce Jacobs' dichotomy below:

Commercial (trader) morals/ethics	Guardian (raider) morals/ethics
Shun force	Shun trading
Come to voluntary agreements	Exert prowess
Be honest	Be obedient and disciplined
Collaborate easily with strangers and aliens	Adhere to tradition
Compete fairly	Respect hierarchy
Respect contracts	Be loyal
Use initiative and enterprise	Take vengeance
Be open to inventiveness and novelty	Deceive for the sake of the task
Be efficient	Make rich use of leisure
Promote comfort and convenience	Be ostentatious
Dissent for the sake of the task	Dispense largesse
Invest for productive purposes	Be exclusive
Be industrious	Show fortitude
Be thrifty	Be fatalistic
Be optimistic	Treasure honor

Clearly "dispense largesse" is a polite phrase for "offer bribes", while "take vengeance" and "treasure honor" seem obsolete today in the Western world, albeit not in the Sicilian Mafia or the Wahabi Muslim worlds. The "commercial" syndrome is closely related to the Protestant "work ethic" and the writings of Benjamin Franklin, while the "guardian (raider) syndrome" is a pretty good characterization of Arthurian knights. Donald Trump is a would-be "guardian" misfit in a "commercial" world.

Ms. Jacob's essential point was that some of the problems of the world today arise because of misguided attempts to apply moral precepts in inappropriate places. It is not appropriate to apply commercial trader bargaining ethics to the civil service, police or army, while "guardian" ethics of Donald Trump do not apply in normal businesses (not even among Wall Street raiders and "activist investors").

---

[1] Niccolo Machiavelli, (1469–1527) was a diplomat, politician, historian, philosopher, humanist and writer of several books, including "The Art of War" (1521) and "The Prince" (1532). He is regarded by some as the father of political science.

But the "spirit of capitalism" is something else. It is closely related to the "Protestant Ethic", of hard work, "time is money", and wealth creation as evidence of moral and "predestination" (Weber 1902). This new ethical idea emerged out of theological argument, starting in the early decades of the sixteenth century. It turns out that the history of modern economics—and of political theory—is largely the story of privatization of common land. Today the distinction between private and public (social) ownership is quite clearly defined in law, at least for tangible goods.

It was not a coincidence that the Protestant Reformation—and its ethic—was soon followed by an explosive growth of long distance trading by wealth-seeking trading companies. Those English companies began as importers of timber, tea, coffee, tobacco, pepper, spices, silk, calico and muslin, for which they paid in codfish, salmon, wool, copper, tin, silver or gold. The first of the English trading companies was the Muscovy Company, chartered 1553. It was really the first multi-national company. Later trading companies engaged in other traffic, including weapons, gunpowder, opium, and slaves. Yet the traders shared an ethic, now taught in economics departments and business schools—especially the Booth school at the University of Chicago—as "rational utility maximization". This ethic is (mistakenly, in my view) confused with fundamental human nature.

It is worth emphasizing, again, that the hierarchical feudal structures of society and government are still reproduced almost exactly in modern joint stock corporations. The first English trading companies, starting in the early sixteenth century, were partnerships owned and financed by wealthy merchants or landowners. The ship's first captains were sometimes partners, but once the routes became familiar, the sailors and crew were hired by the voyage, paid in cash at the end of a voyage, and dismissed. Apart from the captain, the workers got no share of the profits of the voyage, even though they encountered storms, pirates, and diseases *en route*, while the "owners" risked only money and lived comfortably on their estates. It was an extreme version of the "gig" economy now exemplified by the low-cost airline Ryanair and the ride-sharing firm Uber, among others.

The British East India Company was chartered in 1600 by Queen Elizabeth. I. During its first century it engaged only in trade. Later it evolved, first by creating plantations (originally for tea, later for rubber, indigo, and opium[2]).

---

[2] The opium was grown in Bengal for sale to China. The two Opium Wars (1839–42 and 1856–60) were blatant use of superior military power to "open" China to English (and French) traders. The purpose was to profit from an illegal trade, to obtain extraterritorial rights and permanent ownership of Hong Kong. It was a deliberate attempt to increase drug addiction and create demand for smoking opium. The sale of opium for Chinese currency enabled the British merchants to import silk, porcelains, tea and other Chinese luxury products without paying for them with scarce gold or silver.

During the seventeenth century it also became the chief promotor of the British "mercantile policy". One of its officials wrote a book "England's Treasure by Foreign Trade", subtitled "The Rule of Our Treasure". In brief, the rule was *"to sell to strangers more yearly than we consume of them in value"* (Mun 1664). His book—written to guide his son—explains, in detail, all the ways this aim can be accomplished, both by reducing manufacturing and shipping costs and reducing English consumption of imported (Indian) goods. Finally, as commerce merged with government, the East India Company became the overseas arm of Manchester-based manufacturers seeking export markets. It was in that role that the East India Company intentionally destroyed Indian competitors in the cotton cloth (muslin and calico) business.

In the mid-eighteenth century, the East India Company became the nucleus of the British colonial government in India (and the model for colonial expansion by European rivals). The East India Company officially ruled India from 1757 until 1858 (the year of the Indian Mutiny). The trading companies eventually morphed into the British Empire after the Napoleonic Wars and the industrial revolution. At its peak, the East India Company allegedly accounted for half of all world trade in cotton, silk, tea, salt, saltpeter (potassium nitrate), indigo dye and opium. Virtually none of those commodities, except tea, and opium derivatives from Afghanistan, are important in global trade today.

Before "moving on" to the industrial revolution, it makes sense to summarize here—as best I can—the state of the planet in 1750 or so, from an economic and environmental perspective. Briefly, the Earth was not yet over-populated, although famines (and wars) were fairly common. The population was not increasing rapidly, because death rates from infectious diseases, mainly water-borne and especially among children, were still very high. Life expectancy after birth was under 40 years.

What changed in Europe in the eighteenth century, apart from specific inventions and new technologies, was the nature of entrepreneurial capitalism and its (mainly Protestant) adherents. But first there was another change, a changing attitude to knowledge and to discovery.

The Age of Enlightenment: 1585–1789

While rational profit maximizing capitalists were making and investing money, remarkable social visionaries also emerged, somewhat in the tradition of Plato's "Republic". It started with Thomas More's "*Utopia*" (1516), and Francis Bacon's "*New Atlantis*" (published postmortem in 1627). The "age of enlightenment" is traditionally defined as the period after the death of Louis XIV (1715), although some scholars set the beginning back to as early as 1620, to include the writing of "*The New Atlantis*."

I would put it still further back, to 1585, to include writings of Giordano Bruno, Gallileo Gallilei, and members of Walter Raleigh's "school of night" (also known to some as the "school of Atheism"). That was a remarkable group, including Christopher Marlowe, John Dee, Thomas Harriot (who worked for Raleigh and corresponded with Johannes Kepler), "the Wizard Earl" Henry Percy, and others, including several Freemasons.

Newton in England and Leibnitz in Germany were the two most famous pioneers of what became known as "scientific method" derived from astronomy and physics. (Yet laws against witchcraft were still on the books until 1736.) Later, in the seventeenth century, the pioneers of scientific rationality were mostly French "philosophes." They included René Descartes, Francois-Marie Arouet (Voltaire), his wife Emilie du Chatelet, Denis Diderot and Jean le Rond d'Alembert, who jointly initiated the massive 35 volume, *Encyclopedie* (1751–1772). Outside of France, important thinkers were Baruch Spinoza, Isaac Newton, John Locke, David Hume, Jean-Jacques Rousseau, Adam Smith, and Immanuel Kant. There is no need to summarize what they all said—which fills libraries—but to note that their names are still recognizable because, taken together, their words made people think differently about the world and about knowledge and about the act of thinking itself.

Baruch Spinoza, a Dutch Jew, was one of the most important thinkers of the Enlightenment. He was one of the first to say in public—and in print—that the Bible was written by ordinary people, that God does not hear prayers or punish our misdeeds; that God is not an entity outside of nature, that Man is not God's chosen creature, and that there is no afterlife. Spinoza was an intellectual atheist in all respects except that he believed in God. His great work was "*Ethics*" published in 1677.

David Hume, Rene Descartes and Immanuel Kant were among the first to think about the process of thinking itself, focusing especially on the distinction between deductive and inductive reasoning. Hume observed that passion overcomes reason in human behavior (an idea forgotten by neoclassical economists). Hume, like Hobbes, was an empiricist, a skeptic and a naturalist. He was anti-teleological, a Deist, and a precursor of Darwin.

"Humanism" (nothing to do with Hume) emerged as a movement in the eighteenth century. Tawney describes it as follows: *"Humanism is the antithesis ... of materialism. Its essence is simple. It is the belief that the machinery of existence—property and material wealth and industrial organization, and the whole fabric and mechanism of social institutions—is to be regarded as means to an end, and that this end is the growth towards perfection of individual human beings"* (Tawney, 1931 [1952]) p. 84.

Another important new idea of the Enlightenment is that of *progress*, which was first celebrated publically in the great exhibition of 1851 in the Crystal Palace, and after by the Victoria and Albert Museum. Apparently about 6 million people visited the exhibition (a third of the population of the UK at the time) and it made a large profit (£186,000) that was devoted to supporting scientific scholarships. This was the first of the World's Fairs, of which several later imitations are worthy of mention.

For example, Philadelphia's Centennial Exposition marked the peak of steam power technology and celebrated 100 years of US economic and technological progress since the Declaration of Independence. Then came the Columbian Exposition in Chicago (1893–94) celebrating 400 years since Columbus discovered America. That one was also notable for the introduction of alternating current electric power by Westinghouse Electric Co. The "Century of Progress" in Chicago, 1933–34, celebrated the 100th anniversary of the founding of that city. Then came the New York World's fair of 1939–40 which apparently celebrated economic recovery from the Great Depression. Now we have a "World's Fair" somewhere every few years.

Another of the post-Hobbes ideas that emerged during the enlightenment, was the majoritarian idea of "popular sovereignty" as reflected in political discourse around the world today. States are supposed to belong to 'the people', as in "The People's Republic of ..." and government policies supported by privileged elites are legitimized and marketed (like "Brexit") as "the will of the people".

The person who introduced both the idea and the language was Jean-Jacques Rousseau, in his book "*The Social Contract*", published 4 years after his death (1782). It was Rousseau, more than anyone else, who introduced the idea that" The State" is not just a population living in an area. It has an independent identity and even a personality. This idea is implicit in the title of the book.

Another idea that matured during the Enlightenment that has had a powerful effect on events in the "real" world, was the idea of "separation of powers" in government, viz. executive, legislative and judicial. But separation of powers was needed to provide "checks and balances" to constrain a Hobbesian absolutist like the Sun King, Louis IV *("l'etat c'est moi")*. It was an old idea with a new coat. (Aristotle called it "hybrid" government; the Greeks practiced it, as did the Roman Republic.

But separation of powers in practice requires stable and long-lived government institutions, that rarely existed in the distant past. The theory was first discussed in detail and in modern context by Charles-Louis de Secondat, Baron de la Brede et de *Montesquieu* (known only by his last title) in his book "*Spirit of Laws*" (1748). Montesquieu's formulation (see Fig. 1) was explicitly

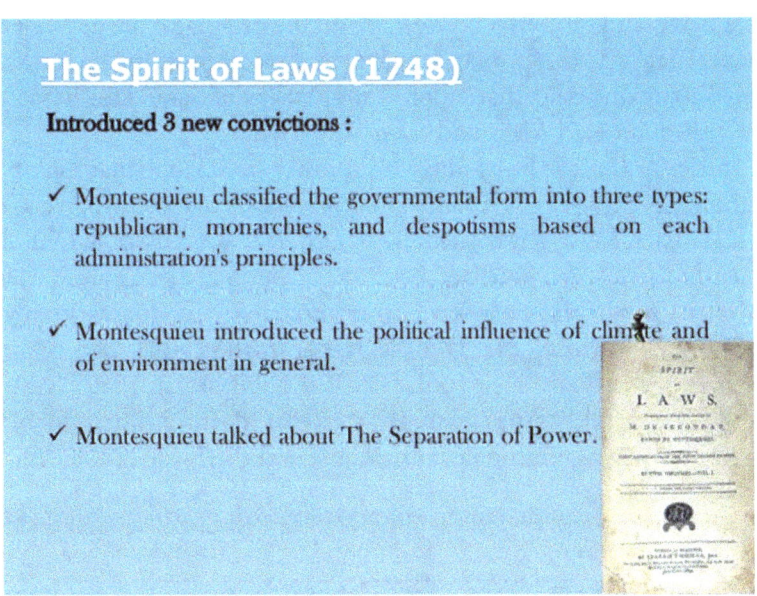

Fig. 1 The book by Montesquieu that inspired the US constitution

incorporated into the US Constitution in 1788. It is important to note that the newly elected, or appointed office-holders in the US today do not swear allegiance to feudal superiors (or their modern equivalents), *but to the constitution itself*. This was a huge innovation.

Montesquieu's work inspired another important political activist and philosopher, Thomas Paine. During his youth in England, Tom Paine was an acquaintance and young protégé of Benjamin Franklin. He arrived in Philadelphia in 1774 (as an immigrant). Two years later he wrote "Common Sense" (1776) a 50 page pamphlet that sold 500,000 copies and inspired the US Declaration of Independence, which was published a few months later.[1] Later, affected by the deteriorating situation in France, Tom Paine wrote "The Rights of Man" (1791), which argued that people have a right to revolt against intolerable oppression. He later wrote "The Age of Reason" (1794), about the place of religion in society. Both books helped to shape the outcome of the French Revolution and the conversion of France into a secular society, the second in the world after the United States of America.

Paine's writings were essential components of what we now call "The Enlightenment" and the Industrial Revolution. What the two revolutions did not change was the social power structure and the deeply rooted assumption that Mankind has rightful dominion, with divine sanction, over all the land, the sea, and all the birds and beasts that live. That assumption is now under challenge. John Locke, in particular, argued that the primary role of government was the protection of private property. In his view, the inalienable rights are life, liberty, and property. Locke was also the originator of the labor theory of value, which was the primary theory of classical political economy, from William Petty (who said that labor is the "father of value and nature is the mother") through Karl Marx and John Stuart Mill.

The stories of Doctor Faust illustrate another sea-change that took place in social thinking between 1620 and 1820 quite well. In the original German folk-tale, Doctor Faustus was a scholar whose story was based on that of a real alchemist and astrologer who lived in northern Germany (possibly Wittenberg, the university town where Martin Luther also lived). The first written version of the story, known as the "Wolfenbüttler manuscript", was written in 1580–87. A "chap-book" for popular sale was published in 1587. The plot, in that book, portrayed Faust as a scholar, very like Martin Luther, full of book-learning but knowing nothing of real people and their way of life.

---

[1] For another example of a book that had a comparable effect, in recent times, Adolf Hitler's *Mein Kampf* sold millions of copies and still inspires neo-Nazis'.

The Age of Enlightenment: 1585–1789

The literary Doctor Faust grew tired of reading dusty manuscripts and became obsessed with gaining other kinds of human knowledge. To further this ambition he made a deal with a powerful demon, Mephistopheles, who showed up conveniently: He agreed to sell his eternal soul to that demon in exchange for 20 years during which he could be a pure libertarian: do anything and go anywhere he wished. During that period Mephistopheles was to be his servant. Here is an English translation of a fragment of the manuscript (*The Wolfenbüttler manuscript of Faust, (1580–1587), at lettersfromthedustbowl.com*):

> "*So uncanny did it become in Faustus' house that none could dwell there. Doctor Faustus himself walked about at night, making revelations unto Wagner as regardeth many secret matters. Passers-by reported seeing his face peering out at the windows. Now this is the end of his quite veritable deeds, tale, Historia and sorcery. From it the students and clerks in particular should learn to fear God, to flee sorcery, conjuration of spirits, and other works of the Devil, not to invite the Devil into their houses, nor to yield unto him in any other way, as Doctor Faustus did, for we have before us here the frightful and horrible example of his pact and death to help us shun such acts and pray to God alone in all matters, love Him with all our heart and with all our soul and with all our strength, defying the Devil with all his following, that we may through Christ be eternally blessed. These things we ask in the name of Christ Jesus our only Lord and Savior. Amen. Amen.*"

It was from this chap-book that Christopher Marlowe wrote his play "*The Tragical History of Doctor Faustus*". In the play, Faust is humanized into a scholar who is dissatisfied with his life among books. But he is so unworldly that he allows himself to be tricked into having a sexual affair with a village girl, Marguerite, thinking it was love. He abandons her, not knowing she was pregnant or what that would mean to her. But as a single mother with no husband, no money, and no family she kills the baby or allows it to die. Faustus learns of this tragedy at the last minute and pleads for her forgiveness and redemption. She is redeemed but for him it is too late. Mephistopheles wins the gamble. Faust goes down to burn forever in Hell.

The message to the reader—and the audience of the play—is perfectly clear: Satan, the anti-Christ, lives among us, always in disguise, and he constantly deceives and tricks us into committing irremediable sins. The message is "Shun such acts and pray to God alone in all matters". It is the same anti-knowledge message conveyed by Adam and Eve's punishment by expulsion from the Garden of Eden, for eating of the fruit of the tree of knowledge. The reader, or listener, is advised to avoid, in particular, scholars and purveyors of alchemy and astrology.

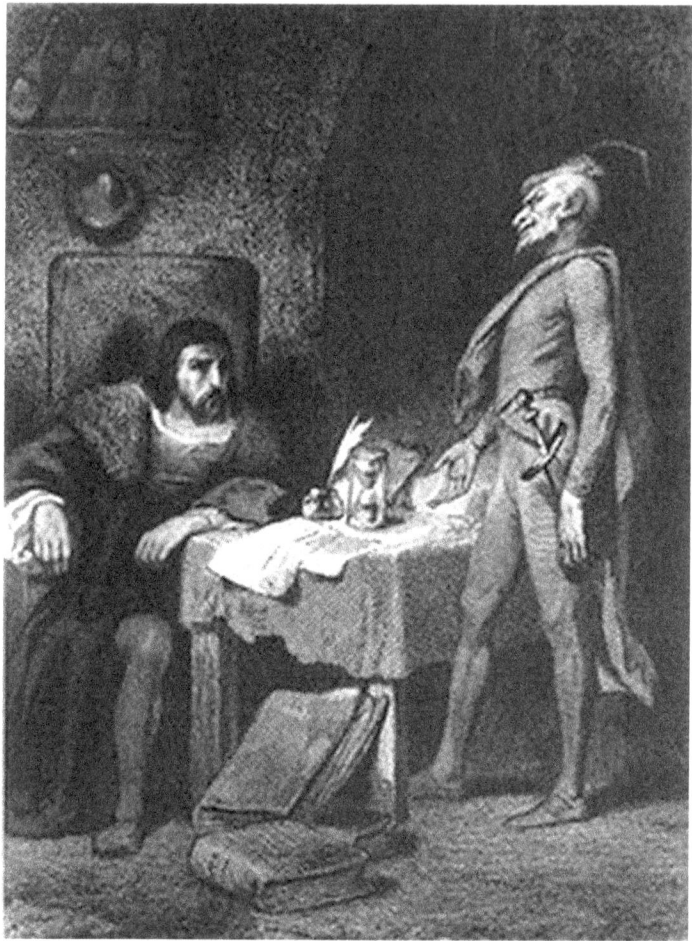

Fig. 2 Mephistopheles offering his "help" to the scholar Faust, Goethe's version of the story; from Allenby, picture source unknown)

Yet, Goethe's (revised) story of the scholar Faust, 150 years later, has a completely different message. As before, Mephistopheles seduces Faust with offers of help (Fig. 2). The human part of the story plot is similar. In Part I Faust is an alchemist seeking a way to convert base metals into gold. It starts with the same human dilemma, a mother (named Gretchen) who is abandoned and who drowns her baby. (Goethe was influenced by an actual trial of such a woman). But the larger message in Part II is radically different. The secret to unlimited wealth is *paper money*, imprinted with the Emperor's seal: Faust and Mephistopheles persuade the Emperor to sign a note: *To whom it may concern, be by these presents known, this note is legal tender for one thousand crowns and is secured by the immense reserves of wealth safely stored underground in our Imperial States*. During the night, Mephistopheles has the note reproduced in all denominations and widely distributed. By the next morning the Emperor has forgotten that he signed such a note. To his surprise these copies are accepted as real money by all the creditors, soldiers and citizens who had been awaiting payment. Faust has unveiled the secret of printing money by borrowing against the future. It was, in Goethe's words "*A continuation of alchemy by other means*". It is not far different from what central banks do today. [1833]1984, Part II. Act V, lines 11,559–11,586 quoted by Allenby (Allenby, 2002).

Goethe's Faust was a very different person from the aging scholar who knew nothing of real life. Goethe's Faust was one who rebuilds the natural environment regardless of cost—and in league with Mephistopheles. Faust's story is usually taken as hubris, extended too far, resulting in nothing but failure and damnation. But that is not the way Goethe wrote it. For one thing, Faust, despite the costs he imposes, is brought to completion, and thus dies, only when he foresees his creation of a world of humans: Mephistopheles, claims Faust's soul at the end, but in vain. Instead, the Heavenly Host, strewing roses, disperses his devils and even makes Mephistopheles himself fall in love—(Lines 11,753–11,800). Faust is saved; as a "*man [who] has gained learning*". He will teach the *Blessed Boys* (lines 12,076–12,084), and (line 12,087) "*will soon be the peer of any angel*". Freedom (Goethe 1833/1984), quoted by Brad Allenby (Allenby, 2002).

There is nothing in Goethe's history about Doctor Faust roasting forever in Hell. Rather, Goethe's Faust is one who "*dares to know*" (*sapere aude,* in old French). As Allenby points out, the Faustian mythology is thus far more ambiguous than commonly realized; it reflects an ambiguity regarding choice, where outcomes are unforeseeable, and the Hippocratic admonition "do no harm" is often not feasible in real life. This ambiguity extends into our own time. It is the ultimate challenge of authenticity in this anthropogenic world.

Money played a little known but important role in the industrial revolution, apart from its use for buying and selling goods. In the Middle Ages money was mostly coins made from copper, silver or gold. All of those metals, after smelting, could be converted directly into coins by the (relatively) simple method of heating and stamping the hot metal into a stone-carved mold. The Swedish military conquests of the sixteenth and seventeenth century were partly financed by the "copper mountain" at Falun. The great pit, over 100 m deep, opened up when the mine collapsed in 1687 (Fig. 3). Most of the copper in Europe today came out of that great Swedish pit.

Fig. 3 The great pit at Falun, in RAÄ buildings database, by Arild Vågen

The Spanish empire in the sixteenth and seventeenth centuries was originally financed by silver from many small mines, especially in Austria. The first big discovery in Europe was the Joachimsthal mine, in Bohemia. The coins made from that silver were called Joachimsthalers. (The word "thaler" later morphed into "dollar".)

The Spanish (Holy Roman) Empire was later financed by treasure looted by the Conquistadores, from the Aztec and Incan civilizations. They took the gold first, but later they found the silver mines of Potosi in Peru (now Bolivia). The Potosi mine was founded in 1587 and was operated by slaves until the nineteenth century.

Neither Sweden nor Spain used their copper (in Sweden's case) or gold and silver (in Spain's case) to finance productive investments. Metal was used primarily for paying mercenary soldiers to fight their dynastic wars and secondarily (in the case of gold) for conspicuous consumption by the aristocracy and the Church and for women's dowries in India. The payments to mercenaries increased the money supply in circulation, without increasing output of goods. Thus, the transfer of gold and silver from the New World to Europe led to more or less continuous monetary inflation. That had other consequences, especially for those countries lacking precious metals, such as England.

In the late fifteenth century, the Portuguese—inspired by Prince Henry the Navigator—discovered the Azores and Brazil. Soon after that, Vasco de Gama rounded the Cape of Good Hope and—to make a long story short—his feat initiated long-distance sea-borne trade. This was a luxury trade that expanded rapidly. The trade with the East Indies (so called) made the coin (i.e. money) shortage in Europe steadily worse, because the commodities the traders wanted to buy could only be obtained by payment in silver or gold.

The gold scarcity did not stop rulers from using gold for bragging purposes. The most extraordinary competitive dazzle contest took place in 1520 at the "summit" meeting (in a village near Calais) commemorating a treaty of friendship between King Henry of England and King Francis I of France. It became known as the "field of the cloth of gold" because both kings were showing off their clothing finery (and that of their courtiers) to symbolize their power and wealth.

The gold shortage was a problem in England, especially because gold coins were continuously exported from Europe to purchase tea, pepper, spices, silk and ceramics (Chinaware) from the Far East. Spain was able to replace its losses of gold and silver from America, for a while, but England could not do that, except by capturing Spanish treasure ships. This activity, initiated by Sir Francis Drake, was called "piracy" by the Spanish and "privateering" by the British. It became a mini-industry for sea captains in the South coast of England during the reign of Queen Elizabeth I.

King Henry VIII's "great dissolution" (1538–40) suppressing the Church of Rome (and seizing its properties) was undoubtedly prompted, in part, by the gold shortage. He confiscated as much as his minions could find of the gold plate and ornaments of the monasteries and churches. And you may ask: where did the gold end up? The answer is: not in commerce. The gold was used mainly for women's dowries in India and China.[2] Figure 4 is a picture of a wealthy Indian woman wearing her dowry. There is no doubt that gold jewelry adds to the attractiveness of a beautiful woman. But the demand for gold dowries has made girl babies into financial burdens to the parents who need to finance dowries for them. This is said to be is the leading cause of female infanticide in India today.

This continual flux of gold from the West to the East finally led, two centuries later, to the so-called "mercantile" policy in England. Everybody who studied US history has heard of that policy, but hardly anybody knows how it worked or what brought it about. The mercantile policy was to sell English manufactured goods from Manchester and Birmingham to its American (and other) colonies in exchange for payment in specie (gold or silver). At the same

---

[2] The gold accumulates (still) in the form of dowries, in India, which are effectively payments from a girl's family to her husband's family, but is kept by the bride as jewelry. It is estimated that there is now 20,000 tons of gold on Indian women, considerably more than in all the central banks around the world. (The financial burden of these dowries, resulting in infanticides of female babies, accounts for the serious female shortage in India—933 girls per 1000 boys).

Fig. 4 The end-use of gold: Indian dowries. fanart.tv/movie/4073 jodhaa-akbar

time the English tried to prevent the colonies from manufacturing goods for themselves. But there was no source of gold or silver in New England or any other part of the original 13 American colonies. The consequence was predictable, but strangely not foreseen in London.

The southern American colonies had two export products, tobacco and cotton, for both of which they got paid in gold coin (or equivalent credit). Hence, they could afford to buy those desirable manufactured goods from the mother country. But the northern American colonies, such as Pennsylvania, New Jersey, New York and New England had almost nothing to export except whale oil and ice (which they collected from ponds in winter and sold to the plantation owners in the Caribbean).[3]

The American colonists—being English themselves—were very unhappy about this treatment, to put it mildly. Imposing a tax (to be paid in gold or silver) on tea imported from India, by the East India Company, was the last straw. The "Boston Tea Party" was a protest in which bales of tea were dumped into Boston harbor by protesters. That protest began the Revolutionary War of Independence. The Colonists finally won the war, after 4 years of hardship,

---

[3] They cut ice from frozen ponds in winter, and shipped it by sea to those who could afford luxuries in the warm places of the Caribbean. They exchanged ice mainly for rum.

with help from France, especially by the Marquis de Lafayette, supported by a hurricane and a slave revolt (in Haiti).

An important invention with a social consequence was Eli Whitney's "cotton gin", patented in 1794. It was widely imitated by planters. The Manchester-based cotton industry received cotton from India but it was also dependent on cotton production by slaves on plantations in the southern states of the US and, as such, a driver of the slave trade (Fig. 5).

US cotton output in 1790 was a mere 3135 bales. Eli Whitney's invention of the cotton gin did not make him wealthy because all the planters utilized his invention without paying any royalties. But it cut costs and manpower requirements dramatically, resulting in a cotton export boom. Five years later, output was 16,719 bales. By 1805 it had multiplied eightfold to 146,000 bales. By 1820 it was 334,000 bales. By 1830 it more than doubled again to 731,000 bales, and by 1850 it was 2,133,000 bales and was still increasing rapidly. It was a boom

Cheap cotton resulted in increased demand for cotton cloth. That, in turn, triggered increased demand for machinery, which also triggered rapid technological progress in cotton manufacturing. This, also resulted in a boom in the slave trade, as well as the creation of many fortunes in southern American States and cities where the cotton was shipped. Fast forward to 1857, there

Fig. 5 A cotton gin in use (operated by slaves). en.wikipedia.org

was a slump in the cotton market (due to over-production) that reduced British enthusiasm for the Confederate cause in the US civil war (1861–65).

Cheaper cotton fiber resulted in increased demand for cotton cloth. That, in turn, triggered increased demand for machinery, which also triggered rapid technological progress in cotton manufacturing. This also resulted in a boom in the slave trade, as well as the creation of many fortunes in southern American States and cities where the cotton was shipped.

A fashion event in 1789 also made a huge difference for the cotton trade. It is encapsulated in this portrait of Marie Antoinette in a simple muslin dress (Fig. 6). That portrait, and the dress pictured in it, created a sudden surge in demand for higher quality cotton fabrics. Calico is another cotton fabric still produced in India. It is much less fine than muslin but softer than denim or canvas. Muslin and calico were important export products of India until Manchester and the East India Company intervened to destroy the Indian cotton industry with British government help. The surge in demand for cotton further ignited the slave trade.

Homo sapiens is not "starting from scratch", from a Hobbesian 'state of nature' where every individual is alone in the world and everyone else is an enemy or potential enemy. Humans rarely choose to live alone. *Homo sapiens*

Fig. 6 Marie Antoinette wearing her muslin dress; portrait by Louise Élisabeth Vigée Le Brun 1783. Photo: Wikipedia/Public Domain

is biologically "hard wired" to live in groups, from couples to families, extended families, to tribes, and communities. We are not yet "hard wired" to live with each other in nations. Families tend to have leaders, as do tribes, cities and nations. In every group there is competition for leadership. Having said that, people in groups and groups within larger groups, economic systems or nations, also need mechanisms for living with each other.

Human history is, in some respects, a history of wars. Wars are not means of living with each other. In far too many cases they result from misunderstandings, failures of communication and unwillingness, on the part of leaders in power, to compromise. Granted, some past wars were necessary and unavoidable. The war to defeat Nazi Germany, between 1939 and 1945 falls into that category. The Korean War was hard to avoid once the North Koreans crossed the line. But journalist I.F. Stone raised questions about its origin and made a case that the United States government manipulated the United Nations, and dragged out the war by sabotaging the peace talks (Stone, 2015). But I (and many others) would argue that Napoleon's invasion of Russia, the Crimean War, The Boer War, First World War and the Vietnam War were quite unnecessary and need never have happened. Luckily—if that is the word—the era of wars fought by large land armies may be over. This may be because of the risk of nuclear war, conventional war morphing into a nuclear war, even if it is very low, is too great to contemplate. Or it may be because there is very little to be gained by "winning" a conventional war, in terms of capturing resources. Wars of territorial conquest are now out of fashion, although the creation of Israel from Palestinian land by the UN seems to be a partial exception.

The US involvement in the ancient religious conflicts in the Middle East may, or may not, be defensible on geo-strategic bounds (depending on viewpoint) but is not about territory per se. It is, to some degree, about access to resources (oil and gas). Here the role of money enters the picture, front and center. Money is the life-blood of all non-primitive economic systems. Without money, or at least an ordinal money-equivalent, there is only barter: e.g. Two coconuts for a chicken. Moreover, barter societies can only survive by being self-sufficient on the basis of local resources. Like the Mayans and Incas, and many tribes in the Amazon, they are helpless to defend themselves against societies with more efficient systems of production and exchange utilizing external energy sources (beyond food and animal feed) and with money.

Unfortunately, as we are told from every media outlet, money cannot buy everything. There are no markets for some things that matter to us, as individuals, as families, as communities, as corporations, and as nations. That statement applies at every level. Money cannot buy happiness, respect, or justice at the individual level. Nor can it buy corporate security or national security.

Money as Wealth

I start with the idea of wealth, in the sense of "what money can buy", leaving out the things it can't buy. Most organizations in human history did not create wealth. In most cases they acquired wealth from nature (by mining, fishing or collecting edible roots, or nuts) or—occasionally—by theft from other groups. Government organizations acquire wealth by taxation, or by (legal) confiscation from persons outside the realm defined by law. The English government obtained a significant share of its revenues by government-sanctioned piracy ("privateering").

Leaders formerly used wealth mainly to acquire and hold power, especially by supporting armies and building castles and fortifications. Churches and monastic orders acquired wealth by donation, either voluntary or coerced. They used most of the wealth thus acquired to feed and house hierarchical superiors and non-working acolytes. The leaders built fortifications, temples and other places of worship, not to mention palaces for themselves. Feudal nobility mostly consumed wealth for self-aggrandizement. But free landowners in favored regions acquired wealth, initially, by retaining economic surplus from high value animals (e.g. furs, elephant tusks, wool) or multi-year agriculture (e.g. mulberry trees for silk-worms, tea plantations, olive orchards, vineyards) and reinvesting it in the land to increase output. Trading profits were reinvested also, to improve roads, protect travelers, improve navigation maps and so on. These people—especially in Italy—used wealth to create more wealth.

At first, the new wealth created by investment was an easy target for the political and military powers that existed at the time. The owners of guns soon acquired the wealth. But gradually governments, and the Church, became increasingly dependent on the wealth creators. As time has passed, both governments and religious organizations have lost influence—and economic power. Much of that power has shifted in the last two centuries to banks and large-scale industrial-commercial organizations, including the so-called "military-industrial complex". The role of law as an enabler of this shift has simultaneously increased.

For thousands of years during the era of hunter-gatherers, the human population grew very slowly. After the start of the Neolithic era (roughly 8000 BCE) aggregate wealth, in the sense of capital goods (e.g. buildings, metal objects, ceramics, infrastructure) grew slightly faster. There were setbacks. In the course of time, agricultural technology (fertilization, seed collection for plant breeding, ploughs, animal husbandry, the use of oxen for pulling the plough, etc.) advanced to the point that agricultural surpluses became possible. This happened first in places with alluvial riverbed soils such as the valleys of the Nile, Tigris-Euphrates, Indus, Ganges, Brahmaputra, Mekong, Yangtze and others. Food (grain) storage followed, along with protective walls and containers.

Trade in commodities also followed. Tree and vine fruits and nuts were gathered and dried or fermented to make wine (stored in clay bottles). Fish were dried and shipped to inland locations. Bread was invented and produced in a hundred shapes and forms. In time the word "bread" became an all-purpose word for money or goods, as in "earning my bread" or "give them bread and circuses". But the biggest innovation of all was the combination of banking and capitalism, in the sense of making money from money.

The three legs of capitalism have been characterized by Robert Solow as self-interest (greed), rationality, and equilibrium. The second and third legs of this trinity are discussed at some length, in the appendix. But many economic philosophers have pondered over the idea of "greed", meaning that enough is never enough. Alfred Marshall wrote, as quoted by his student, Frank Knight "..[it is] *human nature to be more dissatisfied the better off one is*" (Knight and Merrian 1945). His student, George Stigler, went even further "*The chief thing which the common-sense individual wants is not satisfaction for the wants he had, but more and better wants.*"(Stigler 1987).

In recent times the word "sufficiency" has entered the vocabulary of economics, because some observers have seen that increasingly, more of what we already have, is too much. The planet is finite, and while wants may be unlimited, resources available to supply human wants are limited. More to the point, the unlimited wants of the over-fed can make it difficult or impossible to meet the needs of the under-fed.

What is beyond serious doubt is that consumption, driven by "wanting more", is the main driver of market capitalism. A story that may, or may not, be accurate but that conveys the idea, is the following: An economics teacher trying to explain the concept of "sufficiency" asked his students to tell the class how much money would be enough for them. One student at a time responded with a number. The first few answers were modest: "I would be satisfied with a million", said the first. "Two mil" said the next. "Ten mil" said a third. "A Hundred mil" said a bold fourth respondent. The fifth student answered: "all of it". That ended the class discussion, as well as this paragraph.

Observing the activities of many Wall Street characters, that last answer conveys the essence of capitalism: that the winner takes all. This is also the engine of inequality and potentially the fatal weakness of capitalism itself.

Money as a Medium of Exchange

Trade, as barter, preceded money, and both preceded economics. The impulse for trade follows from diversity and specialization. Different regions have different resources. People have different talents and skills. Even in primitive societies some (usually men) were better hunters and fighters, while others (mainly women) were better at recognizing edible roots, seeds or leaves, cooking, sewing, and repairing things, not to mention caring for babies. Anthropologists differ with regard to the division of labor in primitive society, but the answer does not matter much for this book.

Similarly, natural resources are not uniformly distributed. Flat lands in river valleys (like the Nile, the Tigris-Euphrates, the Indus, the Mekong, etc.) are the easiest to cultivate, because they are fertilized by annual floods. But they are not equally good for all crops. Grasslands are good for grazing animals (cattle, sheep, horses), but difficult to cultivate because of deep-rooted perennial grasses and scrub. Forests provide a variety of fruits and nuts from long-lived trees, but forests do not include plants that grow each year from seeds. Many important minerals are found in only a few locations, such as diamonds in South Africa, silver in Mexico, tin in Malaysia and Bolivia, copper in Chile, cobalt in the Congo or platinum group metals in Siberia or South Africa.

It follows from geographical and topographical differences that goods will be less valuable where they are easiest to find and produce, and more valuable where they are scarce. Thus people on islands or near the coast may have plenty of fish and coconuts but not enough milk, cheese or red meat. One of the earliest traded commodities was salt, which is plentiful by the sea or in some deserts, but very scarce in tropical forests of grasslands. The evolution of long-distance trade from salt to spices, to tea, tobacco, cacao, dyes (like cochineal and indigo) and medicinal substances (like quinine) took place because those commodities are scarce and found in very few places.

It is this diversity of natural resources (and specialized knowledge) that creates opportunities for trade. It makes sense for farmers and artisans to specialize in growing or making whatever goods are best suited to particular locations, to transport (or export) those goods to other places with different specialties. The economist's term for this is "*comparative advantage*". David Ricardo's classic example (though he did not use the phrase) was Portugal: He thought that Portugal should specialize in producing wine for export to England, while England should specialize in manufacturing cotton cloth for export to Portugal (and everybody else, including India). The point is that, in principle, trade provides benefits for both parties. In practice, the advantages are usually unequal, and there lies one difficulty.[1]

The above simplified version of trade suggests barter. For instance a man with a fish might exchange it for a cocoanut. Or a Portuguese ship-owner might load a ship with barrels of wine and hope to exchange it for a shipload of bales of cotton cloth. However, trade in the real world doesn't work like that, because the goods in question rarely have the same comparative value in terms of other goods. Putting it another way, a load of cotton goods may be worth more (or less) than a load of wine, if the owners could choose among a variety of other goods to trade for. The solution of the difficulty is for the traders to specify a common *unit of value* that everyone agrees on. That is called a *medium of exchange*, or simply *money*.

There is a new problem when we try to define money. *Is the money just a social convention? Or must it have inherent value itself?* It is tempting to assume that money began as a unit of something with inherent value. The earliest forms of money were items with an immediate value, while being limited in supply. Masai tribesmen in Africa used cattle, Aztecs used cacao beans, Berbers used salt, others have used ivory, coconuts, olive oil, and sheepskins (Weatherford 1997). Gold and silver became money because of their rarity and desirable physical properties. But cowrie shells, used as money in a number of countries, had no use-value and only rarity to recommend them. Yet the earliest records (cuneiform) were tables of debt and credit, which naturally evolved, over time, into monetary balances (Graeber 2011).

---

[1] Consider the silk and tea trades during the Middle Ages. In both cases, there was no actual trade, except of precious metals (silver and gold) for goods, because Europe had nothing else to sell that the East wanted to buy. This one-way flow of silver and gold from Europe to Asia led, inevitably, to the conquest and invasion of the Americas and Africa in the search for precious metals. For a modern perspective, see Magdoff (2000).

Money as a Medium of Exchange

But the major innovation was coinage. For over 2500 years coins have been largely confused with money. The major innovation that enabled the growth of banks and the later rise of capitalism is transferable credit, which doesn't need coins. Its beginning was also an accident. Although the history of money started with gold and silver coins, and only escaped finally from that mind-set in the late twentieth century. The invention of coins did not happen because of a theoretical need. It probably happened, like most historical innovations, by accident. In the Mesopotamian world there are hints that "hoards" of silver objects, of known purity, were exchanged before actual coins. These would have constituted "stores of value" for purposes of trade, but each "hoard" would have had to be weighed and tested for purity. Quite a few "petite hoards" of silver have been discovered, mainly in Israel or Palestine. More are likely to be found if (when) systematic archeological investigations are carried out in the war-torn parts of the Middle East.

It is known that there were copper and bronze coins in circulation in ancient China, but not how it happened (Thierry 2001). In the case of silver and gold, we think we do know. The first silver coins, recognizably as such, were minted by King Pheidon of Argos around 700 BCE. The coins minted by King Croesus of Lydia, impressed with a lion's head, were first minted around 600 BCE. They were made from a silver-gold alloy called "electrum" (70% gold, 30% silver) found naturally in the Pactolus river that flowed through Sardis, the capital of Lydia. (That "river of gold" was also the source of the legend of King Midas, of the golden touch (Bresson 2005).) The Lydian coins had the advantage of known weight and value, under seal of the authority of the state. Their use made Lydia very prosperous as a trade center, for a few years. It had a downside, however: Croesus' wealth soon attracted an invader, King Cyrus the Great, of Persia who conquered Lydia in 546 BCE. He adopted the use of gold coins in the Persian empire (Fig. 1).

Silver and gold coins quickly proliferated throughout the Mediterranean world. But without the Lydian guarantee of quality (based on the use of the alloy electrum), there was always doubt about the metal content. Archimedes famously solved the problem of purity measurement. He was hired to determine whether King Hiero's gold crown had been "diluted" with silver (c. 225 BCE). Archimedes realized—while taking a bath—that the volume of any object could be determined by measuring the volume of water displaced by the object. From the volume and the known density of the metals, he could determine the gold vs. silver content of Heiro's crown. He was so pleased with this discovery that he allegedly cried *"Eureka"* ("I have found it") and ran naked through the streets of Syracuse to tell the world (not shown in Fig. 2 below).

Fig. 1 Persian gold coin c. 490 BCE

Fig. 2 Archimedes thoughtful, by Wikipedia (1620)

Fig. 3 Swedish copper coins from the time of Gustavus Adolfus (mid-seventeenth century)

During the late Middle Ages, money was made from metal from the famous "copper mountain" at Falun (in Sweden), in the form of very large copper coins (Fig. 3). That mountain of copper financed the foreign military adventures of the Swedish kings (especially Gustavus Adolfus, in the seventeenth century) and made Sweden a "Great Power" for a few decades.

It has been estimated that ⅔ of the copper in Europe today came from that single mine at Falun, (see Fig. 3 in chapter "The Age of Enlightenment: 1585–1789") in Sweden. The tin-copper mines in Cornwall that constituted the biggest market for steam engines in the early nineteenth century also provided copper for brass, tin for bronze and for the telegraph wires and the then-burgeoning electrical applications in the nineteenth century. Copper from new mines in Montana, Utah and Arizona arrived in the 1880s, just in time for its use in copper wire for the electrification of industry and homes in the twentieth century.

It is noteworthy that before the middle of the nineteenth century, much of the world's gold and silver came from very few sources. The Spanish "conquistadores" who conquered Central and South America (except Brazil) found a lot of gold already mined and used for religious imagery. But neither the Aztecs nor the Mayas used gold for money. Mostly they used gold to honor the gods, as Fig. 4, below indicates. (The Aztecs used cacao beans as a commercial medium of exchange.) The gold looted by the "Conquistadores" on behalf of the Spanish Crown was sent back to Spain in treasure ships. There, some of it ended up in Churches, or jewelry, but the bulk was used (as coins) to pay the mercenary soldiers hired to fight against the French, the Turks and the Dutch rebels. That gold increased demand for goods and services without financing any increases in productivity. The result was price inflation for many goods.

The Spanish Empire was largely financed by two silver mines, one in Bohemia and the other in Bolivia (Potosi). The Joachimsthal silver mine in northern Bohemia, was discovered in 1519. Its output was made into silver Guldengroschen (later called "Joachimsthalers") that financed the Hapsburg rulers of Austria by paying for their soldiers. In 1546, the Spanish "conquistadores" found a huge silver deposit in what is now northern Bolivia. That deposit became the site of the famous Potosi mine. The town had a population of 200,000 at its peak in the nineteenth century.

Fig. 4 The Muisca raft from which the Zipa offered treasures to the Guatavita goddess in the sacred lake (Meso-America)

During the second half of the sixteenth century, this single mine produced an estimated 60% of all the silver mined in the world. The mine was operated by private entrepreneurs, whose profits were then heavily taxed by the Spanish government. The silver itself was mostly converted into money (coins), most of which was paid to mercenary soldiers in the Spanish Army of Flanders. That silver money ultimately moved on east to China in payment for tea, silk and other trade goods.[2]

Gold, silver and gemstones were the outstanding examples of trade goods that were valuable precisely (and only) because they were both scarce and beautiful, and easily recognizable. Now silver has little scarcity value (even for jewelry) and it takes an expert to determine whether a gemstone is "real" or not. Only gold still qualifies as a "store of value", in itself, although its use as currency is past, and its value for jewelry is, to some extent, culturally determined by its continuing use for the dowries of Indian women.

In short, coins were invented because they were thought to be a "store of value", universally accepted at the time. But not very long after the use of coins became established, it became clear that the valuation of coins

---

[2] It has been calculated that Spain spent more money on its 80-year long war against the Dutch (218 million ducats) than it received from the Potosi silver mine in Bolivia (121 million ducats) during those years (Kennedy, 1989).

themselves was still a problem. In point of fact, the problem was never completely solved until gold and silver coins ceased to be used in the twentieth century. Coins were "official" forms of money within countries where there was a central government that guaranteed their value for domestic trade. But the relative values of different national currencies, with respect to each other (exchange rates) is a continuing problem for international merchants and traders. I discuss this topic in more detail later.

The Swedish copper from Falun also had an important practical use, for roofing and in bronze (for canons). But the gold looted by the Conquistadores in America was only used in Europe for jewelry, religious artifacts or money (as coins). The latter required only the simple physical process of melting, rolling and stamping. The silver (from Potosi) was entirely converted into coins. These coins increased the money supply, at least temporarily, without increasing output of goods. That, alone, caused inflation. Sweden and Spain did not use gold and silver for productive investment. Instead, treasure from America financed conspicuous consumption by the Crown and the Church and (mostly) wars (by paying the wages of mercenaries). This lack of productive investment left Spain behind its rivals when the silver supply from Potosi began to run out in the eighteenth century.

The staple trade goods along the famous "silk road" from China through central Asia to the Black Sea or the Mediterranean, were silk and tea from China (East to West) and horses, in both directions, from central Asia. India also sold tea, spices, dyestuffs, calico and muslin cloth to Europe, during the seventeenth century. However, China and India had little use for either raw materials or manufactured goods from Europe, so the luxury goods purchased by wealthy aristocrats and merchants in Europe had to be paid for in silver or gold coin or bullion.

This imbalance was one of the causes of the European passion for discovering (and stealing) gold, wherever it could be found. It also was a primary cause of the western voyages of discovery and the later military-religious incursions by Portuguese, Spanish, French, Dutch and English explorers in the Americas, Africa and South Asia. The industrial revolution changed the balance of military power, of course, and enabled the creation of the later European empires, that lived by selling manufactured goods (especially cotton textiles and guns), and opium, to the Eastern countries in order to pay for tea, spices, silk and (later) dyestuffs, teakwood, tin and rubber.[3]

---

[3] The so-called "opium war" was a British scheme for creating a market for narcotics in China (based on opium from Afghanistan) in exchange for Chinese currency (silver) to pay for the silk and tea to be sent to Europe. The Chinese government's attempts to stop this trade were met by gunboats.

Of course, technological change also changed the availabilities and scarcities. Silk was once an important trade -good because it depends on silkworms that grow only on mulberry trees, and the silkworm cultivation and silk processing technology was a well-kept secret. It is still a luxury trade -good, but rayon, nylon, orlon, dacron and other synthetics have taken over most of the market for soft, "silky" textiles, replacing cotton, linen and hemp. Alum (potassium-aluminum sulfate) from a mine in Tolfa, Italy, was once a major commodity needed for color-fixation in the wool-dyeing industry; it has long since been replaced by synthetics.

Speaking of dyes, certain colors were once very rare and commanded high prices. Red dye came from the rubia plant in India or from cochineal, an insect. Vermillion, from cinnabar (the ore of mercury) was once used in ceramics and glass, for its red color. Purple dye, from a mollusk found on the shores of the Mediterranean, near Tyre, was so scarce that it was reserved for kings and became known as "royal purple." Now, thanks to petro-chemistry there are no rare colors for dyes.

Aluminum metal was once used for jewellery until the Hall–Heroult process (simultaneously invented in France and the US) made it cheap and ubiquitous. Whale oil was once very valuable for indoor lamps, but it was replaced first by kerosene (illuminating oil), then by gas light and then by electric light (incandescent, fluorescent lights and now by light-emitting diodes (LEDs)). Saltpeter (natural potassium nitrate) was once important for gunpowder, but it was displaced by nitrates from synthetic ammonia. And now gunpowder itself has been replaced by synthetic explosives. All gemstones can be synthesized, and most jewels for watches are synthetic. But, on the other hand, clean air and clean water are now getting scarce.

Skills and knowledge also became more specialized as products became more complex. Early humans made simple clay pots. But artisans over the ages have learned to make beautiful and marvelous porcelain (china-ware) for cooking, eating or simply for decoration. Ming vases required extraordinary skill that takes years to learn. The first woven textiles, based on wool or plant fiber were already a huge step beyond animal skins. But textiles, woven in a cottage from animal hair, are not comparable to a Japanese silk kimono. The first metal products were probably hammered spear points or something comparably simple, whereas Damascus swords or the swords used by Japanese Samurai warriors, are far more difficult to make, requiring very high levels of skill. The same point can be made about performing music on a piano or painting pictures or calculating rocket trajectories or "mining" for bitcoins in a super-computer.

Indeed, complex products are now produced by complex organizations combining many different specialties. This observation led Adam Smith to identify the phenomenon of "division of labor" and to notice that the greater the division of labor, the more efficient a manufacturing process can be (Smith 1776 [2007]).[4] His famous hypothetical example of the pin factory exemplifies the point.

In the nineteenth century, an American industrial engineer named Frederick Winslow Taylor formulated a theory of business organization (known as "Taylorism") based on dividing the overall business into labor functions, each of which can be separately maximized by training and practice. His system included time and motion studies (Taylor 1911). Time studies were pioneered by Taylor himself, while motion studies were perfected by Frank and Lillian Gilbreth. (Their unit of measurement was the 'therblig', Gilbreth spelled backward.) Lillian Gilbreth also wrote a best-seller, "Cheaper by the Dozen" describing her life as a super-mom.

The basic purpose of TAYLORISM was to take control of the manufacturing process away from the skilled machinist or other skilled workers. Henry Ford was an early follower of Taylor's scientific management methodology for productivity improvement. Mass production and the moving assembly-line were logical implications of Taylorism. Extreme versions of this mechanistic approach are boring and dehumanizing (as even Adam Smith pointed out). Modern factories usually prefer some version of team-work, where members of the team work together, learn each other's functions and become partially substitutable for each other.

---

[4] To be fair, the notion of division of labor has been discussed by many philosophers before Adam Smith, including Plato, Xenophon, Ibn Khaldon, Wm Petty, David Hume, Bernard de Mandeville and Henri-Louis Duhamel de Monceau.

Credit and Banking

The religious Order of the Knights of the Temple of Solomon (known as "Templars") were dedicated European (mostly French) soldier-monks (about 10% were knights) with the original mission of protecting pilgrims en route to sites in the Holy Land (and later, as protectors of the Christian enclave in Jerusalem). As part of this mission, they began to accept the responsibility of temporarily storing the deeds of properties belonging to traveling pilgrims. These valuables were kept in the Templar's guarded castles ("commanderies") assuring relative safety. For this purpose the Templars provided their clients with encrypted descriptions of the assets in their keeping. (It is no accident that bank buildings, as recently as the twentieth century, tended to look like fortresses.)

After 1150 CE—with the Pope's permission—the Templars began offering *letters of credit* (LOCs), enabling depositors to draw funds at other Templar commanderies. This worked because the Templars—all of whom were celibate monks who had eschewed personal wealth—were trusted. That trust made banking possible. It also made the order wealthy. This was partly because the Templars charged fees for their services, and partly because they kept the title deeds to the properties of the crusading knight-errants. Some of those knights died in battle in the Holy Land and never returned to their homes, making the Templars (and not their heirs) richer in terms of land ownership.

To avoid calling the fees "interest" (and risk a charge of usury), the Templars called it "rent" for the use of the properties for which they held the deeds. By the beginning of the fourteenth century, the Templars mission had shifted largely away from guarding pilgrim's lives to guarding their property. Moreover, they found that they could make loans to others (such as kings) based on the value of the properties they were guarding. In fact, they found that they could make more loans—in the form of paper letters of credit—than the assets they actually had in hand as reserves. This "fractional reserve" lending practice was very profitable.

One unintended consequence was the end of the Templars, themselves, mentioned earlier in connection with the rise of nation-states and monarchies. But a major legacy of the Knights of the Temple of Solomon was fractional reserve banking, as practiced today in most banks of the non-Islamic world. After the Templar's disappearance, there was a gap that needed to be filled. The gap was filled in the Italian city-states, notably Florence, Genoa, and Venice. The service they provided (transferable credit) was needed to support long distance trade. The early banks were family affairs (Bardi, Peruzzi, Accaluoli, Medici). Those family "super companies" were mainly traders in olive oil, wine and (in the case of the Bardi) woolen cloth. Their banking business depended—as with the Templars—on trust. When trust failed, as when King Edward defaulted, banks collapsed.

Meanwhile, the letters of credit LOCs, originally guaranteed by the Templars, began to be traded among travelers, especially merchants, in the fourteenth century in Italy. LOCs became a form of money. Thence, their guarantors (who also physically protected the valuables of depositors) became banks, albeit under another name. In 1308 King Edward II of England was forced to dismiss a distrusted councilor, Piers Gaveston, and appoint 21 barons as "ordainers" to supervise and approve his "ordinances". The ordainers wrote a document, called the *Ordinances of 1311*. That document gave Parliament control over the administration, prevented the King from going to war or making land grants without Parliament's approval, and provided for a means of monitoring the ordinances. Later kings revoked the ordinances, but still had to take the ordinances of 1311 seriously. Their past existence gave Parliament money-power over the King and (incidentally) established the law, as a profession, in England.

In 1310, international finance played a role in English political history. King Edward II had borrowed £22,000 (a very large sum for the time) from the Frescobaldi bank in Florence. He did it to finance a war against Robert the Bruce of Scotland. That war was not successful and King Edward's Parliament did not want to pay for it. The default had a very bad effect on Italian bankers, as may be imagined.

The first noteworthy financial collapse was the simultaneous failure of the three "super companies" of Florence, in 1344–45. This was partly due to internal Florentine power struggles, and partly due to the repudiation of war loans by Edward II of England. According to one source, Edward owed

900,000 gold florins (£135,000) to the Bardi and 600,000 florins (£90,000) to the Peruzzi, which he could not pay. (Peruzzi records suggest that King Edward did repay some of his loans in cash and some in wool.) But the effect of the losses on Florence was devastating.

The Medici family moved its HQ to Florence in 1397, and engaged primarily in trade, especially wool cloth. They expanded to Venice in 1402. At their peak, the Medicis had branch banks in Lyon, Geneva, Avignon, Bruges and London. The Medici's financed the Roman Curia and later intermarried with French royalty. Catherine Medici (1519–89) was the "mother of three kings". Another Medici (Marie) was married to the next king, Henri IV of France, and later acted as regent to the young future King Louis XIII. The Medici branch banks were financially independent of each other, in order to be protected from the failure of any one branch. Other Florentine banking families included the Cherchi, Gondi, Salviati, Scali and Strozzi.

The Italian banks gained an edge in the sixteenth century thanks to the invention (in 1494) of *double-entry bookkeeping* by Luca Pacioli. The key to this accounting invention is to record every transaction in two separate accounts, usually a "cash" account and an "asset" (credit) account. In this system a loan is recorded as a *subtraction* from the cash account and a corresponding *addition* to the credit account. The two accounts have to balance at all times. In fact, what accountants do to this day is called "balancing" and corporate financial statements are called "balance sheets". Keeping the two accounts physically separate (and maintained by different clerks) also constituted an effective protection against fraud.

The Fuggers and Welsers, both of Augsburg, Bavaria, joined the fraternity of international bankers in the fifteenth century. The Fuggers later took over from the Medici and financed the rise of the Hapsburgs to power in Austria-Hungary. Jakob Fugger was elevated to the nobility in the Holy Roman Empire in 1511. But the Fugger bank was put out of business when Philip II of Spain suspended interest payments in 1575.

The other German bank, also originally from Augsburg, was the Welser bank. They, along with the Fuggers financed the Holy Roman Emperor Charles V (father of Philip II). The Welsers maintained trading posts in Antwerp, Lyon, Venice, Seville, Madrid, Lisbon and Santo Domingo. In 1528 they also acquired a province in Venezuela, as security for a large loan to Charles V. They tried for a while to convert this province into "little Venice." However, they got into trouble for excessive rapacity and lost the province in 1556. Luckily for them, Philippine Welser married Archduke Ferdinand II of Austria. The Welsers moved their HQ from Augsburg to Nuremberg late in the sixteenth century. Banking was a good way for commoners to "buy in" to

Fig. 1 The Welser expedition to look for gold in Venezuela

the nobility (and even to royalty). Figure 1 below recalls the Welser "army" of adventurers, on its way to search for *El Dorado*. It didn't exist, but the gold-thirsty adventurers left their mark.

Banking spread to other cities including Antwerp (Berenberg–Gossler–Seyler), Geneva (Bordier), Zurich (Hottinger), Siena (Chigi) in the sixteenth century and to Amsterdam (Hope), Cologne (Oppenheim), Hamburg (Schroder, Warburg), Frankfurt (Metzler, Rothschild) and London (Coutts, Hoare) in the seventeenth century. *Each of these banks issued its own coinage.* By the sixteenth century, international monetary exchange was too important to be disorganized and haphazard. A new species of financial intermediates had to be invented.

National currencies created by sovereign mints gradually displaced private localized forms of money controlled by individual banks. But in the seventeenth century the national currencies were continually being "cried up" or "cried down" by the sovereign of a country. The purpose was to change the market value of the currency (and to devalue sovereign debts). This was a form of indirect taxation on merchants, and silver coins were constantly being debased by rulers. The 40% devaluation by Henry VIII was notorious.

Consequently the market value of silver bullion kept rising in comparison with the nominal silver content of the coins. In other words, extraordinary as it seems today, it was increasingly profitable, for a while, to collect silver coins and melt them down to recover the silver content.

One of the most famous phrases in economic history is "bad money drives out good money". What it means is that, if there are two kinds of coins in circulation, one of which is suspected to be "watered" or counterfeit, merchants (and customers) will always choose to pay their bills with the counterfeit coins, keeping the others in reserve. This is called "Gresham's Law", because Sir Thomas Gresham, who was the agent of the English Crown in Antwerp and later co-founded the London Exchange, was the first to state it this way. However, the phenomenon was understood earlier by a number of earlier authors, including Aristophanes, in his play "The Frogs" (fifth century BCE).[1] In 1551, Gresham was working for the Regency Council of the Protectors during the short reign of the boy King Edward VI. (Edward, who was crowned at age 9 and died at age 15, was the sickly son of Henry VIII. He died in 1553, probably of tuberculosis.)

The problem facing the Council of Protectors in England (1551) was that the pound sterling had lost half its value since 1544, from 26 Flemish shillings per pound to only 13 Flemish shillings per pound. The decline was continuing because the "financial market" (i.e. international bankers) did not trust English money any more (Martin 2014) p. 110. This was a big problem for England because during the last years of Henry VIII's administration and that of Queen Mary who was married to, and influenced by, King Philip I of Spain, the English government had borrowed a lot of money in Flemish shillings to finance war. Now the lenders wanted to be repaid in that currency. But the courtiers of the Regency Council in England blamed the bankers for deception. According to lawyer William Cecil (then a Minister, later chief advisor to Queen Elizabeth I) the Italian bankers *go to and fro and serve all princes at once ... work what they list and lick the fat from our beards*" (ibid).

To stabilize the pound sterling, Gresham first proposed a "stabilization fund" to allow him to take counter-party positions (i.e. to buy sterling whenever the bankers sold). But the Regency Council lost confidence in the plan and stopped the fund after a few weeks. Then Gresham came up with another idea: the Regency Council should confiscate the private foreign currency reserves of all the English merchants in Antwerp, calling this a "loan to the crown".

---

[1] Other authors who used the phrase include Nicholas Copernicus Nicole Oresme (c. 1350) and the Mamluk Al Maqrizi.

This coup was approved and implemented. It immediately raised the value of the pound, making Gresham a hero to the regency government (but a villain to his fellow expatriate merchants). Naturally the merchants objected bitterly, but to no avail. But in the long run the monetary discipline imposed by the currency market was effective. In fact, Queen Elizabeth I, who succeeded to the English throne in 1558 (with Sir William Cecil as her chief advisor) was extremely conservative in financial matters throughout her reign. (Her policy of moving herself and her entourage from one great house to another great house every summer was largely a money-saving device.)

Because of the constantly changing relative values of coinage, the inner circle of international bankers, all known and trusted by each other, invented their own valuation system. It was called *ecús de marc*, a scheme to formalize the changing relative value of national currencies (Martin 2014). They met at regular (3 month) intervals starting in 1555, in Lyons, France to fix exchange rates for the next period. Inter-bank payments, in terms of (LOCs) from other banks took place at those fairs. On the third day of each fair, inter-bank payments were made and new credit balances for depositors were established.

In 1609, the Bank of Amsterdam (Wisselbank) was created by the City of Amsterdam. This bank, like others, created its own money. It accepted gold and silver bullion and all foreign coins at real value (based on metal content), melted them down to recover the metal, and re-issued its own coinage known as "bank money". The bank charged a small fee for the coinage service. The net value of a deposit was then kept on the books of the bank. (Clients had to pay the bank to store their coins.) All bills worth more than 600 guilders had to be paid in bank money. This made Amsterdam, in effect, the central bank for the world for the next six decades, until the creation of the Riksbank of Stockholm in 1668.

Here I need to mention an extraordinary polymath, scientist, philosopher and economic theorist, Sir William Petty. He wrote an important *Treatise on Taxes and Contributions* (published in 1662) and several other writings published posthumously. He was the first to use the phrase "*laissez faire*" (meaning "let it be, or leave it alone"). He was the first to calculate national income, national wealth, money supply and the velocity of money. He estimated in 1662 that the average income per person in England to be £6 13s 4d, and the population to be 6 million. This yields a national income of £40 million. He also calculated that the value of capital stocks (land, ships, estates and housing) to be £250 million, of which the returns ("flows") were £15 million. On the same basis he calculated the aggregate value of the labor stock to be

£417 million, treating the £40 million national income as "flows". He also calculated the money supply (silver and gold coinage) to be £6 million.

Petty was a believer in the "classical" (Aristotelian) idea that labor and capital are the "mother and father" of wealth. His writings had a great influence on others, especially Richard Cantillon, the French "physiocrats", Smith, Say and (much later) Keynes. Richard Cantillon, one of the first French Physiocrats, is worth remembering for another reason. In 1730 he wrote "*The Land is the Source or Matter from whence all Wealth is produced. The Labour of man is the Form which produces it; and Wealth in itself is nothing but the Maintenance, Conveniences, and Superfluities of Life*". These words come from his "Essai sur la Nature du Commerce in General". It is one of the earliest writings that reflect that essential fact that wealth is derived from nature. Energy was not understood at the time, but it was understood that land provided something that was essential to the growth of crops.

The next (or possibly the first) central bank in the modern sense was the Riksbank in Stockholm, Sweden, formed in 1668. This bank was given the authority to fix the reserve requirements for other banks in the country, and to store all the gold and silver deposits centrally. Other European central banks followed suit in the following decades.

The Civil War in England during the 1640s was the first serious revolt by "the commons" against the arbitrary power ("divine right") traditionally exercised by kings. In fact, the republican government that was installed by Oliver Cromwell's victorious "new model army" did not take root for various reasons. The Stuart dynasty was restored in 1660 with Charles II on the throne. However, the "restoration" (of the monarchy) was very limited in terms of royal power. King Charles II no longer had sovereign control over the money supply (or of much else). Parliament was in charge. His role was mainly ceremonial, not much different from the monarchs of today.

When Charles II's brother, James II, succeeded to the English throne in 1685 he tried to restore Roman Catholicism as the State Religion (with support from Louis XIV in France). There was no support in England for such a change. King James was expelled from the country. The "Glorious Revolution" followed very quickly and a Dutch King, William of Orange, was literally imported, with an army, in 1588. But King William arrived with his own agenda. It was to create a Protestant alliance to fight against France, mainly to preserve the independence of the Netherlands.

The English-Dutch alliance won that war, notably at the Battle of Blenheim, thanks to the brilliant generalship of the first Churchill. (He was elevated to become the Duke of Marlborough as his reward for that victory.) But the

Fig. 2 The "Old Lady of Threadneedle Street" (Bank of England) founded 1694

English Crown had been deeply in debt even before the war, and the war taxes—as usual—produced less revenue than the costs. So the deficit got worse. Finally, the English Crown lost its creditworthiness and could find no foreign lenders. By 1694, the financial situation of the government had become really critical.

The solution, proposed by a man named William Paterson, was simple. A private bank, to be called the Bank of England, was to be created. It would receive the right to issue its own money *in the name of the Sovereign*. It was created and it still exists (see Fig. 2). In return, the bank agreed to lend some of that money to the Crown. This "great settlement" occurred in 1694. But the problem of coinage had not yet been solved. Most of the coins in circulation were then considerably under-weight, thanks to a variety of schemes to remove silver by clipping, filing or shearing, without making the coin unrecognizable. Something had to be done about the degraded coinage.

A former Treasury official named William Lowndes was asked to make proposals. Lowndes proposal in 1695 was to reduce the weight of the silver in the new coins by 20%. It quickly turned out that the price of bullion had risen so far that a deeper reduction in silver content was needed. The revised proposal was put to Parliament. All the practical "money-men" thought it was the right thing to do.

However, the philosopher John Locke, chief theorist of constitutional government—who was greatly admired and extremely influential—was invited by Parliament to comment on the Lowndes proposal. Locke thunderously opposed Lowndes on the ground that

> "*Silver is the Instrument and Measure of Commerce in all the civilized and trading parts of the world ... and the measure of commerce by its quantity, which is the Measure also of its intrinsick value. ... a pound is a reference to a definite weight of silver*".

and Locke was sarcastic abut Loundes' ideas

> ... "*fanciful justification that there was some metaphysical plane on which a coin would retain its 'value' despite losing 20 percent of its weight.*"

And so on (Martin 2014). Locke may not have been the first of the intrinsic value believers (sometimes known as 'gold-bugs') but he was by far the most influential, at least back then.

The bottom line was (unfortunately) that John Locke got his way. Parliament proposed to collect all the light-weight silver coins in circulation, melt them down and reissue full-weight silver coins. A deadline was set. Citizens with under-weight silver coins were told to use them to pay taxes, or buy government bonds, to get credit for the face value of the coins they turned in. Otherwise they would receive only the partial value based on the actual silver content.

Sadly, a lot of the citizens didn't get the message or didn't believe it. When the coins in circulation were collected, the face values added up to £4.5 million but the silver content was only worth £2.5 million in new full-weight coins (Martin 2014). A great many people cried "foul". Moreover, all new silver coins rapidly disappeared, because the value of the silver in Antwerp or Amsterdam was even higher than the value of the full weight English coins. But, thanks to Locke, the government set a fixed price for gold and a fixed relationship between gold and silver. That was the origin of the "gold standard".

By the middle of the eighteenth century, the British parliamentary government—no longer beholden to the Crown—needed taxes to pay the interest it owed to the Bank of England. The need for gold as a monetary base for the pound sterling was acute. This drove the initiation of the so-called "mercantile" policy, under which England exported manufactured goods to its colonies, only on English ships, in exchange for payment in gold. In 1764 Parliament passed the Currency Act forbidding the American Colonies from printing their own money and requiring all taxes to be paid in gold or silver. (This was because of the continuous drain of silver to China and gold to India, to pay for the Indian tea that was becoming popular in England, as noted already).

The mercantile policy was primarily a tool to achieve favorable trade balances with its colonies, as recommended originally by Thomas Mun (Mun 1645 [1946]). As we know now, and should have been obvious, the mercantile policy combined with the Currency Act of 1764, and the Navigation and Trade Acts were ruinous for the American colonies. They were especially hard on New England, which had no gold or silver mines and had to compete in export markets with the mother country itself. The Townshend Acts (to tax the American Colonies in order to pay the cost of governors and judges) contributed to the disaster. This foolish and self-destructive combination was the primary underlying motivation for the *Declaration of Independence* and the American Revolution. The American Revolution was followed in short order by the French Revolution, the Napoleonic era and—concurrently—the industrial revolution.

Money as Gold

The notion that money should have "intrinsic value" was very much alive in 1800. In fact, Locke's view of what money is, and what it isn't (noted earlier) was supported by most other economists in the nineteenth century. Jean-Baptist Say said in his *Treatise on Political Economy*, that money is a commodity (meaning that it has intrinsic value), and that its value is determined by supply and demand, like other commodities (Say 1803 [1821]). John Stuart Mill used almost the same words in his *Principles of Political Economy* (Mill 1848). Mill was more devoted to the labor theory of value than was Say.

Adoption of the gold standard by the Bank of England (BoE) was a logical implication of this notion. Paper money was printed by the BoE and later by other central banks, only to the extent that it was backed up by a stock of gold such that (in principle) the pound or dollar could be exchanged for gold. Of course there never has been enough gold on hand in every bank to make this possible. But only since 1933 has that fundamental reality been seriously faced and challenged. In 1971 the gold standard was formally abandoned by the US government, during Richard Nixon's presidency. So-called "gold-bugs" and extreme conservatives still insist that this decision was wrong, because they have always assumed that easy money was bad policy and that inflation would erupt if the money supply were to grow "too fast". But their continuing predictions of catastrophic inflation have been consistently wrong, as inflationary pressures in the US have been declining since the early 1980s.

The 1914–1918 World War changed everything, of course. But it had an especially great impact on global money, and the role of gold. The causes of the war itself were not primarily financial and were outside the realm of this book. However the money supply played a role. In 1913, Britain had a money supply (currency in circulation and bank deposits) of $5 billion ($104 billion $_{2011}$). This was backed by $800 million ($17 billion$_{2011}$) in gold, of which only $150 million ($3.1 billion$_{2011}$) was in the vaults of the central bank (BoE). The rest was in the form of coins or bullion held in other British banks. During the war, the money supply increased to $12 billion ($200$_{2011}$) whereas the gold supply did not change. Britain spent $50 billion ($85 billion $_{2011}$) on the war itself, largely by selling its overseas investments to private American investors. Domestic prices increased by a factor of two and a half. The problem for the BoE in the post-war era was what to do about the gold standard and the reserve status of the pound sterling (Ahamed 2009).

France spent $30 billion ($500 billion $_{2011}$) on fighting the war, of which only 5% was paid for directly by French taxes. The rest was borrowed. Half was covered by war bonds sold domestically and $10 billion ($170 billion $_{2011}$) came from loans by the US banks or British banks. The Rothschilds were centrally involved in all this. The loans left an unpaid gap of $2.5 billion that was filled by printing new money. As a result, the currency in circulation in France tripled whereas the quantity of goods produced did not, and prices rose accordingly (Ahamed 2009).

At the beginning of the Great War, a gold-backed Deutschmark was fixed at 4.1 marks to the US dollar, but by the end of the war it was 8.91 marks per dollar. By the end of 1919, it was 47 marks per dollar and by November 1921, it was 300 marks per dollar. The "London ultimatum" of May 1921 insisted that the reparations be paid in gold or dollars at the rate of 2 billion gold marks per year, plus 26% of German exports. By 1921 Germany was collecting just 10 marks in taxes for every 100 marks of expenditures, and had lost all but $120 million of its $1 billion pre-war gold reserves. The reparations to France and England (agreed under the Treaty of Versailles) accounted for a third of the German budget deficit during those 2 years.

The Treaty of Versailles, which ended the war in Europe, was designed (by the victors) to permanently disarm Germany. It was extremely vengeful, partly because of the gross asymmetry of the war damage. German armies had caused enormous physical damage in France and the Low countries. But Germany itself was physically undamaged. The French, especially, demanded financial reparations, as did the Russians. However, the reparations demanded (in gold) were unpayable, since Germany did not have a large gold reserve. President Wilson of the U.S. and John Maynard Keynes (the best economic thinker of the twentieth century) both denounced the treaty at the outset, but they were ignored.

The first result of British-French attempts to enforce the reparations policy of the Treaty of Versailles was that Germany printed Deutschmarks and used them to buy gold from the US (which allowed this exchange by treaty). The actual result of this policy, of course, was that the price of gold in Deutschmarks went up and the value of the Deutschmark, in dollars, went down. The result was hyper-inflation of the Deutschmark in 1921–23. The hyper-inflation wiped out the bank savings of the German middle class, while creating opportunities for unscrupulous capitalists with access to "hard" currency. (Some were Jewish, or thought to be Jewish, which was unfortunate for the European Jews, later.) That, in turn, energized both the communists and the "national socialists" (Nazis).

But, while the domestic budget deficit could have been dealt with by borrowing from the German treasury (as the Japanese and Chinese do now), the foreign debt had to be paid in gold or hard currency. This left the Reichsbank of the Weimar Republic no choice (in the opinion of its president-for-life, von Havenstein) but to issue un-backed paper money to purchase gold from the US. The hyper-inflation took the value of the gold Reichsmark from 4.2 to the dollar in 1914 to 11 trillion to the dollar in 1924 as shown in Figs. 1 and 2 below.

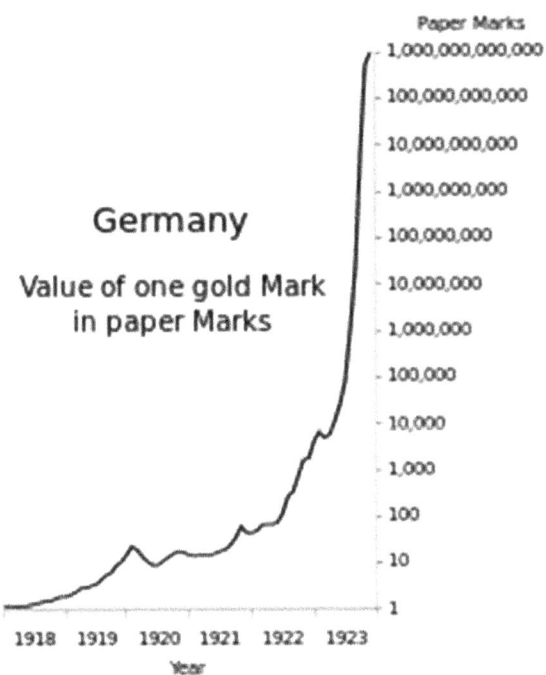

Fig. 1 The hyper-inflation in Germany from 1921 to 1923, after WW I

Fig. 2 German children playing with stacks of currency in 1923

But Finance Minister Gustave Stresemann, and his newly created commissioner Hjalmar Schacht, did an "end-run" around the Reichsbank. What they did was to create a new currency, the *Renten Mark*, backed by land rather than gold. The trick was to limit the total number of Renten Marks in circulation to 2.4 billion, worth about $600 million. The new currency circulated alongside the old one. The exchange rate was fixed by Schacht at 1 trillion old Reichsmarks to 1 Rentenmark. Amazingly, the old currency meanwhile regained some value (from 11 trillion to the dollar, back to 4.2 trillion to the dollar.) Almost miraculously, the currency had been stabilized at a stroke.

At that point, the German Government was able to pay off all of its internal debts, which were valued at $30 billion (in old Deutschmarks) for 190 million in Rentenmarks or $45 million in terms of the old currency (Ahamed 2009 p.189). This coup was cheered by the industrialists and financiers. It was the device that permitted economic recovery to occur. However, that massive wealth destruction simultaneously wiped out the savings of the middle class, that were in houses and government bonds. The Stresemann government also cut costs and carried out a number of other fiscal reforms. All of this resulted in a balanced budget by the summer of 1924. The Weimar Republic was saved, for the moment, but at enormous cost to middle class savers who had patriotically bought government bonds. The German middle class was impoverished at a stroke, while the wealthy, whose wealth was in shares of industrial firms or land, were the winners. This gross inequity seriously undermined social coherence in Germany. To survive, the government needed scapegoats. As usual, the Jews were blamed.

Back to the question facing the Bank of England: What was to be done about the gold supply? In 1925 Britain, under Chancellor of the Exchequer Winston Churchill, the production and use of gold coins was ceased, although Britain still retained the so-called "gold bullion standard" for its money supply. But, influenced by imperialist pride and BoE chief Montagu Norman, Churchill foolishly returned to the pre-war gold standard at the symbolic pre-war exchange rate of $4.86 dollars to the pound, up from $4.40. That exchange rate was a matter of pride for the British upper crust, who were hoping the regain for London the financial primacy is had had before the war. As in Germany, the traveling upper class enjoyed the "strong pound" while the working class were faced with price increases on imported wheat and beef from Argentina and Australia.

Churchill's fixed exchange rate was much higher than the economic fundamentals could support. Exports were weak and domestic agriculture struggled against imports. Indeed, the social consequences of the weak post-war recovery led to great discontent among the blue collar workers of Britain. There was a General Strike in 1927 that united all the unions and nearly led to a social revolution. (That self-destructive policy, driven by symbolism, has a modern equivalent, namely "Brexit", albeit Brexit is not about social change.)

The German reparations called for in the Treaty of Versailles were still unpayable. There were years of fruitless negotiations among the parties, including the US banks, as to ways and means to resolve the problem. The US had spent $30 billion on the war, of which $10 billion was in the form of bank loans to other countries. That money was borrowed from private US banks (mainly the ones in New York), not from the US government. They expected repayment in full. For the US taxpayer WW I was already expensive: It increased the national debt from just over $1 billion in 1914 to $25.5 billion at the close in 1918. But, for the next 11 years, the federal budget was in surplus, reducing the wartime debt by 36%. In 1930 the US public debt was down to $16 billion.

The outcome of the first set of negotiations about war debts was the so-called Dawes Plan of 1924, named for Charles Dawes, a banker and President Harding's first Director of the Budget. (He became Vice President, under Coolidge, later that year.) Under the Dawes plan, German reparations payments were reduced and restructured. The French occupation of the Rhineland was ended. The Reichsbank was to be supervised by foreign creditors. The US banks were then able to shift most of their risk to the public, through the sale of Dawes bonds (which were supposedly backed by gold, but were not).

The Dawes plan, in operation between 1924 and 1929, enabled German firms and institutions (e.g. towns) to issue bonds, in turn. Those bonds were sold to middle-class US citizens. They were designed to outrank (be "senior" to) the official reparation payments, hence making them first in line for repayment. This neat trick enabled Germany to import a great deal of capital during those years to cover both its domestic deficit and its annual reparations payments (Ritschl 2012). The Dawes bonds, backed by war debt, paying 7% with 25 year maturity, was a great financial success story, for the financiers, at first. However, the consequence was that German domestic debt increased substantially during those years after 1924, culminating in a financial crisis. That crisis killed the first Dawes plan.

In 1929 another commission, led by Morgan banker Owen D. Young, cut the German reparation still further and stretched them over 59 years. A new set of bonds was issued to help Germany pay off its Dawes bond debts. The other significant outcome of the Young Commission was the creation of the Bank for International Settlements (BIS) in 1930. The BIS's ostensible purpose, when it was created, was to funnel German reparations payments back to France and England and thence, indirectly, back to the US lenders. (BIS later helped the Nazi government during WW II. It was resurrected to manage the Marshall Plan after WW II and now acts as a kind of Central Bank for Central Banks.) The Dawes bond investors were left to twist in the wind. Most lost their money. It was another hit on the middle class, leaving the wealthy unscathed.

In 1927 Britain, France and Germany sent a joint delegation to Washington to try to persuade the US Federal Reserve to cut the "rediscount rate" for foreign borrowers (i.e. themselves). (The rediscount rate is the rate for overnight loans by Federal Reserve to the commercial banks, which determines all other rates.) The Fed also bought US government bonds from the private sector (a policy now known as "quantitative easing"), thus providing the banks with both cash in hand and low interest rates. *Many of them used this cash, either themselves or by their depositors, to purchase common stocks on margin.* The US stock market bubble of 1925–29 was driven at first by the postwar industrial growth, The Federal Reserve's Index of production rose from 67 in 1921 to 126 in June 1929. But towards the peak in 1929 it was driven partly by inappropriate "quantitative easing" by the US treasury.

The financial wealth lost during the 1929 stock-market crash and the following years, was enormous in historical terms. The story is well-known and need not be re-told here. See, for example, (Galbraith 1954). But that loss of wealth was accompanied by a massive debt write-off in the form of bankruptcies. The debt write-off had the effect of increasing private credit, thus allowing room for new lending. This enabled the economic recovery that began in the late 1930s and continued for the next half century.

The historical "store of value" theory—that money is coins and that coins ought to have intrinsic value—has not yet disappeared completely, but it is now a cult of limited appeal. It began to die in the aftermath of WW I when war reparations in gold were imposed on Germany by the Treaty of Versailles. That had the effect of impoverishing the German middle class, many of whom voted for the National Socialists (Nazis) instead of the conservatives they would normally have voted for, in 1933.

The role of gold as currency was also sharply diminished by President Roosevelt in 1933. But its role in international trade was strengthened, if anything, by the Bretton Woods Agreement of 1944, which required each country to maintain its currency value with respect to gold (within 1% of a nominal rate). Bretton Woods also introduced the International Monetary Fund (IMF) as enforcer and agent for bridging temporary financial imbalances. The US dollar—backed by gold—was the effective international reserve currency. Gold as money was finally killed by inflation and by President Nixon's forced closure of the "gold window" (which allowed foreign governments to buy US gold at a fixed price). This happened in 1971, when Switzerland, Germany and France chose to buy gold instead of buying depreciated dollars. Nixon's action thus broke the legally fixed price of $35 per oz. Since the 1930s, paper money backed by governments (through their national banks) have replaced all but small coins as currency for day-to-day transactions. This has happened, in essentially all countries.

Money as Printed Paper

The "new alchemy" of Dr. Faustus in Part II of Goethe's version of the story was just a recapitulation of Chinese history. Paper money backed by the central government was first invented in China during the Song dynasty (960–1279) (see Fig. 1). This money was probably the first example of transferable credit valid throughout a country. (Note the complex lithography, which was to prevent counterfeiting.)

The next several dynasties after Song, from Yuan to Qing, all issued paper money. Usually each Emperor issued his own. The Yuan dynasty further developed the paper money system used in the Song dynasty and formulated regulations for the use of paper money. It stipulates in detail the date of issuance, the name given to the reign, and prescribed punishments for counterfeiters, ranging from a fine to death, in an extreme case. These regulations made the paper money system of the Yuan dynasty more mature. The Ming dynasty, which followed, continued to promote paper money vigorously. The Ming Dynasty issued a version of paper money, called Da Ming Bao Chao (see Fig. 1). There was a serious attempt, during that period, to rely exclusively on paper money; but later China reverted to using both paper money and coins.

Money today is no longer a store of intrinsic value. It is simply a mechanism for transferring credit. As mentioned earlier, China has used paper money for the last 1000 years but Europe was slow to follow, because of its long obsession with silver and gold coins as currency. In the nineteenth century, Europe coins were first supplemented, and only gradually replaced, by pieces of printed paper, marked with numbers and decorative lithographic patterns designed to identify the issuing bank and make them recognizable and difficult to counterfeit. Those pieces of paper—formerly issued by banks,

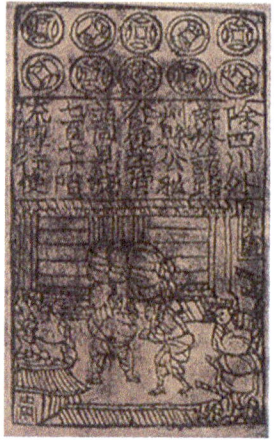  

Jiaozi (Song era)   Zhi Yuan Bao Chao (Yuan era)   Da Ming Bao Chao (Ming era)

Fig. 1 Chinese paper money

now issued by governments—were universal as recently as 2000. They are now being replaced, in many places, by electronic bank transfers.

Yet paper money, too, seems to be on the way into the trash-basket of history. Starting in the early nineteenth century, people wrote "checks" on a bank account and the checks moved (by messenger or mail) to a "clearinghouse". In the "clearinghouse" money was manually subtracted from one account and added to another account—as required by the rules of double-entry bookkeeping. Deposits were often made in a different bank, adjusted by a market "exchange rate" between the currencies of different banks. The "clearinghouse" function was originally an independent business, but after 1880 or so it has been taken over by the banks themselves. The use of electromechanical calculators, and cash registers for merchants began around that time.

Another important change, made possible by the computers, was a new invention, the credit card. This first came into use in 1950. At first, it was just an accounting gimmick. The Diners Club card was the first. It allowed peripatetic businessmen to eat and entertain at 27 prestigious restaurants around the country, with bills processed by the bank and presented monthly. The American Express green card first appeared in 1958. It offered the same service but on a wider scale. Bank of AmeriCard was introduced in California a year later. It later became VISA and is now the most widely used card in the world.

In 1966 Master Card was introduced by the City Bank of New York, also initially for customers with good credit ratings. It was advertised as the "everything card" with a new feature: *revolving credit*. For the first time it was not necessary to pay off the balance each month. This was the first true credit card. The credit card was (is) identified by a magnetically imprinted Personal

Identification Number (PIN) number, known only to the owner of the card. Point-of-sale machines, mostly by National Cash Register (NCR), would "read" the card and match it with the owner's PIN. Thereafter the accounts were processed electronically, and the card-owner gets a monthly bill. If the bill is paid immediately, there is usually no interest charge. But if the bill is not paid immediately, interest charges accumulate. Interest rates on credit-card debt are quite high—much higher than home mortgage interest rates, for instance.

With credit cards, came credit ratings. Credit cards were issued at first only to people known to the bank, with high personal "credit ratings", based on their history of borrowing and repaying loans. Now the rating system has become largely automated. Everyone using a credit card has a credit rating, and the ratings are stored in centralized data storage entities. (The largest credit-rating storage agency, called Equifax, had its supposedly secure data base "breached" by hackers in the summer of 2017. The data included names, addresses, phone numbers, and passwords for over 140 million Americans and 500,000 people in the UK. Nobody knows, yet, what has happened to that information.)

More to the point, in the 1970s, banks started issuing credit cards to young people without prior credit experience, relying on statistical evidence that older card-users almost always repaid their loans. This became a major source of income for the issuing banks, because interest rates on unsecured credit card debts could be (and still are) much higher than interest rates on secured loans, such as mortgages or auto loans. Credit cards have become a convenient way to borrow small sums from the issuing bank. Most people have several cards. Some use them indiscriminately, as though the thought of repayment never occurs to them. Credit cards constitute a major source of profit for banks. US credit card debt is now very large, exceeded only by home mortgages and student loans.

By 1975 or so, all of the arithmetic of money transfer was done electronically in the banks by 'main-frame' computers. Today, your bank account is a block of information stored in a bank's computer, or in the "cloud". The inter-bank transfers are made electronically via the "information super-highway" that Al Gore talked about. Money transfers are now just electronic information flows.

The latest innovation is electronic money transfers for consumer purchases, using "PayPal", "ApplePay" and other such systems. China is once again taking the lead in this technology. In Fig. 2 below, Chinese calligraphy brushes are being sold using patterns on Chinese smartphones.

The last few paragraphs deal with consumer banking, whereas most of the "action" and most of the risks are elsewhere. The 1933 Glass–Steagall law separated investment banks from commercial banks, to prevent the use of consumer deposits as backing for risky financial activities, such as selling

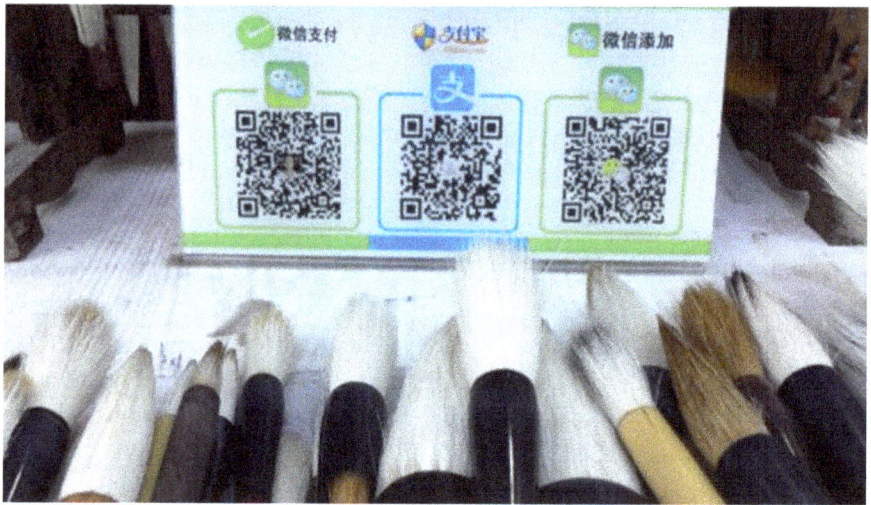

Fig. 2 Chinese calligraphy brushes for sale with pay-by-smartphone (Evelyn Cheng | CNBC-TV)

derivatives, managing mergers and acquisitions, stock-broking and wealth management. But in 1998 the Glass–Steagall law was revoked and US banks (and most foreign banks) are now 'universal' in the sense that they can engaged in almost any kind of financial activity, subject to regulation.

The main point here is that banks today do not only create 'ordinary' money by lending, they also create new kinds of exchangeable objects, such as shares of stock, "junk" bonds, mortgages and "derivatives". Mortgage-based bonds were 'invented' in the 1980s and quickly became the basis for a boom in housing finance. Instead of lending money directly to home-owners (as savings banks used to do) commercial banks now create bonds from packages of mortgages that can be sold to insurance companies or pension funds. This idea was generalized to the creation of "collateralized debt obligations" or CDOs, based on other kinds of debt. These also can be traded. The notorious credit default swaps (CDS) were invented in 1994 as a kind of insurance against adverse financial conditions, such as interest rate changes. They became one of the most profitable financial products of American International Group (AIG), the world's biggest insurance company. Now we have other innovations, such as currency swaps, repurchase agreements, hedge funds and other specialized funds. Since all of these are fungible assets, they can also be used as collateral for loans. This makes it possible to create pyramids of tradeable assets based on other assets, such as the Investment Trusts that were created in the late 1920s and that contributed to the 1929 crash.

Capitalism vs. Socialism: A Conflict of Ideas

Some animals are territorial, others are not. Some are social, other are not. Humans appear to be among the most sociable and territorial of all. The primary resources to be acquired by primitive tribes, apart from defensible locations, are territory in which to hunt, arable land suitable for crops, tamed animals (cattle, sheep, goats, pigs, etc.), forests and access to fresh water (preferably with fish). Land is the source of all biological and mineral resources; it can be "improved" by forest clearing, wild animal capture and breeding, agriculture, irrigation.

If the territory available to a tribe is inadequate (or is perceived to be inadequate) to support the population, the next step would be to increase the territory under control by conquest. Throughout human history, the most effective leaders are remembered by their conquests. We make statues of "great" conquerors (usually on horseback) from Cyrus the Great to Alexander the Great, to Julius Ceasar, to Attila the Hun, to Genghis Khan, to Tamerlane, Charlemagne, to Peter the Great, Napoleon Bonaparte, Frederick the Great, Catherine the Great, … the names are familiar. But what were their actual achievements apart from winning battles?

Conquest of land for agricultural colonization is now out of style. Adolf Hitler's stated goal of acquiring (by conquest) "Lebensraum" for German-Aryan colonists in Russia was a "pipe dream" from the start. Conquest of territory, today, is accomplished by multi-national companies, by means of brands, not armies. Coca Cola and Kleenex are exemples. Tribes and countries once fought over land to colonize. Later they fought over land for gold, diamonds, oil or fish. Countries still compete, but normally within limits short of extreme violence. (The continuing violence in the Middle East has its origins partly in competition for oil but mainly in Religious history.)

Until the eighteenth century, "property" really meant land. Land ownership back then was tightly linked to the feudal prerogatives of monarchy, where all the land was owned by the king and the lands of the princes, dukes, counts and barons were "fiefs" held in fiefdom to the king. The *enclosure movement* was, effectively, a conversion of "common" (public) land into "private property." John Locke—an ardent opponent of the feudal notion of absolute monarchy—argued nevertheless that private property was a "natural right that pre-existed monarchy" (Locke 1960 [1698]). Adam Smith, on the contrary, viewed property as an "acquired right" (Smith, 1776 [2007] #4712).

Within a settled country there is competition among towns, and among organized groups, some seeking wealth, others with different objectives. The profit-seeking organizations compete for investments and for customers or clients, or they may cooperate by sharing markets (contrary to current "antitrust" laws in most capitalist countries). But other organizations (e.g. police, public health, education, communication) that provide services to the whole population do not work well in a competitive profit-seeking environment.

The nature of reality—as matter (not "spirit")—had become an important topic (known as "materialism") for philosophers since the Greek "atomists". Discussion was encouraged by scientific discoveries of Copernicus, Galileo, Newton and others. René Descartes (1596–1650) was perhaps the first to see the human body as a kind of mechanism ("Discourse on Method"). In 1748, French doctor and philosopher, La Mettrie, introduced the first materialistic definition of the human soul in a book "L'Homme Machine".

Between 1780 and 1840 there was an intellectual revolution in philosophy, centered in Germany. It began with the publication of Immanuel Kant's "Critique of Pure Reason", which focused attention on the difference between a "thing in itself" that is inherently un-knowable (e.g. God, or the soul), versus the appearance (to the human observer of that thing [Kant 1781]). The perception of a thing depends upon human modes of receiving information, processing and organization of knowledge. They were also inspired by the industrial revolution in England and its "empiricist" philosophers (Hobbes, Locke, Hume, et al) and the French Revolution.

But Kant's work led to a number of dualisms and contradictions that resulted in several decades of academic dispute among a group of philosophers, calling themselves "idealists" including Johann Fichte, Friedrich Jacobi, Friedrich Schelling, Gotlob Ernst Schulze, and Jeorg Wilhelm Friedrich Hegel. The two major critiques of idealism were Pierre-Joseph Proudhon's "The Philosophy of Poverty", while Marx countered with "The Poverty of Philosophy". Proudhon was the first to call himself an anarchist, and was the author of many quotes, including "property is theft". Hegel had the greatest impact on later generations, especially on the ideas of Karl Marx and "Das Kapital" (Marx, 1867). Looking back, it was the idealists, especially Hegel, who first realized that knowledge has a social character and that the creation and spread of knowledge also moves society. It was mainly Hegel who conceived of political development as a dialectical process of social interaction or discourse, among groups.

During the nineteenth century, Karl Marx and Friedrich Engels borrowed the notion of Hegelian dialectics, (without their "idealist" aspects) and developed "dialectical materialism", a methodology to explain societal change. Marxists, atheists to the core, argued that "society"—as a generality—is a misnomer, because it depends on the mode of production and has different meanings in different times and places. But they did assume that, once the capitalists are gone, the educated elite could mold the old institutions to serve the new egalitarian society. The hope, (expressed by Marx in the *Communist Manifesto)* was that the intelligentsia (i.e. the communist party) would create a society such that "*from each according to his ability, to each according to his need*". As we know, this never happened when communists got power in the Soviet Union, in 1918–19, although (to be fair) the new regime was given little opportunity to put its philosophical ideas into practice.

The idea that industrial workers and miners might organize themselves into "unions" for common benefit emerged at the same time, partly due to the spread of primary education, and partly due to the growth of large enterprises with many workers. Mines were among the earliest. The first proto-union—to provide mutual insurance against sickness, old age and death—was the keelmen (lightermen) in Newcastle, England. This happened in 1699.

On the other hand, "utopian socialism", as envisaged by Christian reformers like Robert Owen, Charles Fourier, J.P. Godin, and Adriano Olivetti, has no philosophical antecedents. It is fundamentally based on the antithesis of "greed is good". It assumes that greed is immoral, by definition, and that society based on cooperation can exist and that it matters. (Remember that Catholic doctrine forbade lending money at interest, calling it *usury*, a sin.) Utopian socialism was based on the idea of sharing and reciprocity, associated with teachings of Jesus Christ (as well as Confucius, Buddha and others). They, too, have been disappointed. It is clear enough from the results of a number of idealistic social experiments in the past that the elimination of greed in commerce was a vain hope.

In this context, the main feature (and arguably, the crucial virtue) of theoretical capitalism, as understood by academic economists, is that it engages selfish (amoral) motives in competitive activities that may yield indirect benefits to society as a whole (Adam Smith's "invisible hand"). For Adam

Smith—who was a Deist—the invisible hand was literally a divine intervention. Today that view is obsolete. But the supposed societal benefits of the "invisible hand"—mainly trade—have always been secondary, at best. Gordon Gekko, the central character in writer Thomas Wolfe's novel "The Bonfire of the vanities"—made into the movie "Wall Street"—put it simply and memorably: "*Greed works; Greed is Good*".

Most economic theorists since Adam Smith have insisted that self-interest (greed) is the dominant driver in every part of society, allowing no role for cooperation, still less altruism. In effect, they deny that "society" has any existence as an entity in itself other than providing a framework for markets. They doubt that altruism actually exists in the real world. (They argue that actions that appear to be altruistic are just another form of selfishness; in their view, Buddha, Jesus Christ, St. Francis of Assisi and Mother Theresa were either masochists, or they were accumulating "brownie points" for their expected post-mortem sojourns in paradise.)

Game theory began with a book entitled "The Theory of Games and Economic Behavior" by the mathematician John von Neuman and economist Oskar Morgenstern in 1944. It was significantly extended by John Forbes Nash, in 1950 through 1953 (Nash 1950, 1951, 1953). Nash showed that there can be equilibrium outcomes when each player takes into account the strategies of other players. Evolutionary game theory (EGT) is the application of game theory to evolving populations in biology. It originated in 1973 when John Maynard Smith and George R. Price formalized contests as strategies, and identified mathematical criteria that can be used to predict the results of competing strategies (Maynard-Smith & Price 1973). Evolutionary game theory focuses on the dynamics of competition among strategies and the frequency of the competing strategies in the population.

Evolutionary game theory has helped to explain the existence of altruistic behavior in Darwinian evolution. Social scientists and psychologists—most notably Amos Tversky and Daniel Kahneman—have pioneered research on the psychology of prediction and probability judgment; later they worked together to develop "prospect theory", which aims to explain apparently irrational human economic choices. I have learned from a reviewer of this MS that behavioral economists have found, in controlled laboratory experiments and field tests, using positron emissions tomography, that we humans possess altruistic-specific receptors in our brains (Kent Kitgaard).

The distribution of wealth in the world today, is grossly unfair and unjust (the richest 0.1% own 50% of all global wealth according to assertions in the media. This is absolutely inconsistent with equality of opportunity. But redistribution of wealth by the state, whether by land-reform, taxation or confiscation—especially if enforced by faceless bureaucrats—is also unacceptable to most people, even to the near-poor. From this perspective, growth is the only answer to the plight of the poor. "A rising tide lifts all ships" is the slogan. So the question is: Does the tide really rise for all? If so, what makes it rise? If not, why not?

The standard answer, by mainstream economists and political conservatives, is that the key to economic growth is re-investment of profits from trade in new products and new markets. Government regulation is widely regarded as a "headwind" whereas "animal spirits" and competition (based on greed) are the "tailwinds" that keep the game going. Innovation is supposed to be a driver of investment behavior. True, competition is a powerful spur to innovation, whereas cooperation is not as effective as an incentive to innovate. But, extreme inequality, being inconsistent with equality of opportunity, is also a headwind.

Without cooperation we would, indeed, have Hobbes hypothetical "state of nature ... red of tooth and claw." In that world, the competition never ends until the final confrontation, when the winner takes all and becomes a dictator (Gilgamesh) or an absolute monarch: *L'Etat c'est moi*" (Louis XIV). Unconstrained individualism—Hobbes' "state of nature"—leads eventually to unrestrained domination by a few oligarchs or only one. Of course Hobbes' defense of a strong king was an argument based on false assumptions about the "state of nature", which was a hunter-gatherer society that certainly depended on cooperation and may, actually have been quite egalitarian.

Pure unconstrained capitalism is essentially a zero-sum game. It is "zero-sum" because the game is about money accumulation (monetary wealth) and the amount of money on the table, during any given round of bets, is fixed by the amount of gold in reserve, or by the Central Bank. Gains by winners as the game progresses are matched by losses by other players. As the game goes on—under conventional rules—there is a systematic transfer of wealth from the losers to the winners. This happens because, in the real (economic) game, the winners have significant advantages that improve their chances.

The most important of those advantages is the law of compound interest: those with money in the bank earning interest, or in long term investments, get richer automatically, as long as the economy itself grows. How could anybody with a few dollars to invest in nineteenth century America fail to make money? "To he who hath, it shall be given" says the Bible (Matthew 13:1), reflecting reality. Apart from making money from their own, or access to (other people's) money, the rich can afford better education, better sources of information, more and better contacts, better lawyers, better financial advisors, better credit ratings, and so forth.

The transfer of wealth, from losers to winners—given a fixed pool to start with—means that, as time goes on, there is less and less opportunity for the rest of the players. In the real economy it means that more and more people are "out of the game" in the sense of having no economic surplus with which to "bet" i.e. to save, invest, or speculate. The fraction of the US population that is now "out of the game", in this sense, is about 60% (Dalio 2019). The "winners" not only capture the wealth that is at risk and "up for grabs", they also keep reducing the opportunity for others to compete—play the game—as they become richer. Less opportunity means less hope. The game ends when one player finally captures all of the wealth that is "on the table". In the real world, things don't usually get quite that far without a violent revolution led by those losers who have been deprived of opportunity, or never had any.

The only way to keep the "game" going—to avoid riots in the streets and a *coup d'etat* of some sort—is to do one of two things. Either keep adding money to the "pot" by creating new wealth, or redistribute the existing wealth from the winners. The creation of new wealth is usually by discovery, invention or innovation. It is largely really outside the domain of government, though government activity in several domains, especially aviation and space programs, has been critical. Government support for scientific research in general, and R&D in certain fields, such as health care, also creates wealth.

The redistribution of existing wealth can be accomplished directly by land reform, taxation or by confiscation, as in the Russian Revolution of 1917. Or It can be accomplished indirectly by devaluing existing wealth. The devaluation and revaluation of existing assets can be a consequence of technological change, as happened when coke from coal replaced (and devalued) charcoal, when steamboats replaced sailing schooners, and steam-powered harvesters replaced men with scythes and mules. Model T's devalued horses and buggies, and a 1000 other examples can be cited. The devaluations resulting from progress in lighting technology are particularly illuminating: They started when incandescent electric lights replaced candles (and devalued sperm whales), tungsten filaments replaced carbon filaments, fluorescent lights replaced tungsten filaments and LEDs replaced fluorescent lights. This kind of perpetual revaluation occurs constantly, if less dramatically, in the stock markets of the world.

Finally, it is clear that inflation is also redistributive. Making the dollar (or other currency) (worth less in buying power), takes wealth from the *rentier* class—people living on fixed incomes from endowments or pension funds—and giving it to debtors, like mortgage-holders whose debts are being paid off in current dollars.

Economic growth occurs when the revaluations of certain assets resulting from discoveries or innovations out-weigh the devaluations elsewhere. In the 1920s, unskilled farm labor was devalued by mechanized cotton gins, harvesters and tractors on farms, but was revalued as industrial labor in factories in Detroit, Birmingham, and Stuttgart. However, "digitalization" accompanied by "globalization" seems to be the process of devaluing and revaluing low-skilled human labor in the OECD countries and revaluing the same skills in developing countries. Shipyards and steel mills in the OECD countries have been devalued by more modern facilities with cheaper labor in Asia. Will the consequent asset revaluations from cheaper ships outweigh these devaluations? And who will gain and lose? That remains to be seen.

Another open question is how the apparent conflict between competition and cooperation will be resolved. In the economic game, the winners tend to be defenders of the *status quo ante*, meaning they favor unrestrained competition and free movement of capital and labor. Yet groups—whether organized as partnerships or corporations—readily out-compete individuals in all kinds of enterprise, thanks to economies of scale and "division of labor". Both kinds of organization depend on cooperation. Obviously both competition and cooperation have a role to play. Human society, in the large, is not unregulated Hobbesian competition between unattached individuals and tribes. Indeed, there is evidence that the role of cooperation *vis a vis* competition in society is actually increasing over time (Axelrod 1984; Axelrod and Dion 1988).

Modern governments in most countries, are built on "voting rights", a consequence of popular sovereignty. In Athens, or the Roman Republic, only landowners had the right to vote. In time, the right to vote has been extended in stages to all male citizens, then to freedmen (former slaves) and finally to women. As noted above, rural landowners—leftovers from feudal society—controlled most governments by virtue of the voting laws, before the twentieth century. They used their power in their class interests, including the privatization of "common property". For instance, the "enclosure movement" in England (one of Tawney's favorite topics) took common village grazing land for the benefit of large wool-exporting landowners, thus destroying the livelihoods of many marginal farmers and turning them into "vagabonds". Vagabonds were not even allowed to move into the towns. This was capitalism at its worst.

True, since the industrial revolution, many of the vagabonds found jobs in the urban textile mills, at near starvation wages. Only as the nineteenth century wore on, did some of them finally benefit from the union movement. This enabled industrial workers to bargain on more nearly even terms, with the land-owning capitalists. By the beginning of the twentieth century, the political side of the union movement ("Marxism") had its own economic theory, and considerable momentum. It peaked with the Bolshevik revolution in Russia, but was tainted subsequently by the horrors of Stalinism. The non-Marxist union movement in the U.S. peaked in 1935.

A Conflict of Ideas, Continued

My father, in his memoirs, recalled an episode in 1928 when he was a graduate student at Columbia University, living in International House. A fellow student tried hard to recruit him into the Communist Party, on the grounds that only the Communist International (Comintern) had any hope of changing all the wrongs in the world (due to capitalism). The supposed advantage of the Comintern, as a change agent, was its unity of purpose and lack of factional disputes. All members of the Party would automatically support whatever the leadership decided. My father was not convinced by this argument, but many liberals were. Anti-Communism, especially after WW II in the USA, became a tool for weakening the industrial unions for the benefit of the owners.

Capitalism is usually interpreted, today, as the system in which the production of goods (and services) is largely left to private enterprise and "free markets," with minimal interference by government *except* in defense of property rights and enforcement of contracts, no matter how one-sided or unfair. The difference between rights and equity—in the sense of "fairness"—has only recently emerged in a few countries as Courts of Equity. Capitalism, today, is mostly the defense of free trade and free movement of capital (but not labor), "globalization," and the war against protectionism. These are important issues but peripheral to the scope of this book.

In the past two centuries, especially since the rise of "socialism" in the nineteenth century, socialism has been widely mis-understood in the upper classes as a collectivist system opposed to private property. That was true for the communards in Paris, the Russian Bolsheviks (and the Comintern). Both were tiny groups, originally. The one that took power (in Russia) equated communism with "dictatorship of the proletariat" and gave the idea a very bad name. But most of the workers' organizations that emerged in the nineteenth century, like the Industrial Workers of the World (IWW), were mainly seeking the right to organize unions for bargaining purposes.

Here, I cannot resist another quote from Mr. Keynes, this one in 1933:

The decadent but individualistic capitalism, in the hands of which we found ourselves after the war, is not a success. It is not intelligent, it is not beautiful, it is not just, it is not virtuous, and it does not deliver the goods…

(in sympathy with those who argue, like Mahatma Gandhi)

… that goods should be homespun whenever it is reasonably and conveniently possible" (Keynes 1933).

There is plenty of evidence that the "hub and spoke" model of industrial globalization, as applied to the supply chain for major industries, such as autos, aircraft, or smart phones, is remarkably similar to the British colonization policy that led to the US Declaration of Independence. The policy continued in the nineteenth century, when African and Asian colonies were converted (by gunboats and the gold standard), into suppliers of cheap raw materials and markets for British manufactured goods. But that subject is also outside the scope of this book.

Capitalism today is theoretically about "competitiveness" and maximizing rewards for risk-taking. According to ideologues, like Ayn Rand and Milton Friedman, capitalism openly celebrates unfettered economic competition and "free markets"—even where the winners take all the marbles—relying on Adam Smith's "invisible hand" to spread the benefits to society as a whole (Friedman 1962; Rand 1964, 1967). For most capitalists, the "'name of the game" is minimizing regulatory burdens, cutting taxes, and avoiding competition to the greatest extent possible. Warren Buffet calls it "widening the moat."

Socialism, today, is about poverty amelioration, provision of standardized social (and other) services—especially health care—and equality of opportunity. The dichotomy between capitalism and socialism will continue, of course, because it is truly fundamental. Marxist socialism deplores, and opposes on principle, the resulting social inequality, and misallocation of resources toward production for profit, as opposed to production to satisfy human needs (Marx 1867a, 1867b [1946]). To a considerable extent, each side of this divide mischaracterizes the core beliefs of the other. Few, if any, socialists would advocate equalization of incomes even though they object to the gross inequality of

A Conflict of Ideas, Continued

today's income distribution. On the other hand, many social democrats would insist that "free markets" are preferable to central planning because they actually lead to better—if grossly unequal—outcomes for all.

Economic theory owes a lot to banking, as already explained in previous chapters. One of the most interesting characters in capitalist history was an Irish-French banker named Richard Cantillon, a follower of Richard Petty, mentioned earlier. In 1716, Cantillon bought a bank (with borrowed funds) and invested a lot in John Law's Mississippi Company. Then he sold out before the top, making him very wealthy. Then—long before Goldman-Sachs—he apparently loaned the proceeds to other investors at high interest rates, payable in Amsterdam or London. His investors pursued him with lawsuits (and may have murdered him by burning his house, in 1734). Cantillon wrote about his ideas (in French) "*Essai sur la Nature du Commerce en Generale*" in 1730, undoubtedly based in part on lessons learned during the Mississippi Bubble (see Fig. 1).

Cantillon's book may have been inspired by William Potter's earlier (1650) monetarist book "The Key to Wealth." Potter attributed economic growth to plentiful money supply (in terms of banknotes). Cantillon's book was far more advanced. It was arguably the first "complete" theory of economics (well before the physiocrats or Adam Smith). His manuscript was translated into English and published in 1755 and later translated into Spanish. His book elaborated John Locke's quantity theory of money, explained prices as the balance of supply and demand (including the interest rate as the balance of supply and demand of loanable funds) and explained interest as the cost of compensating lenders for the risk of default (bankruptcy) by borrowers.

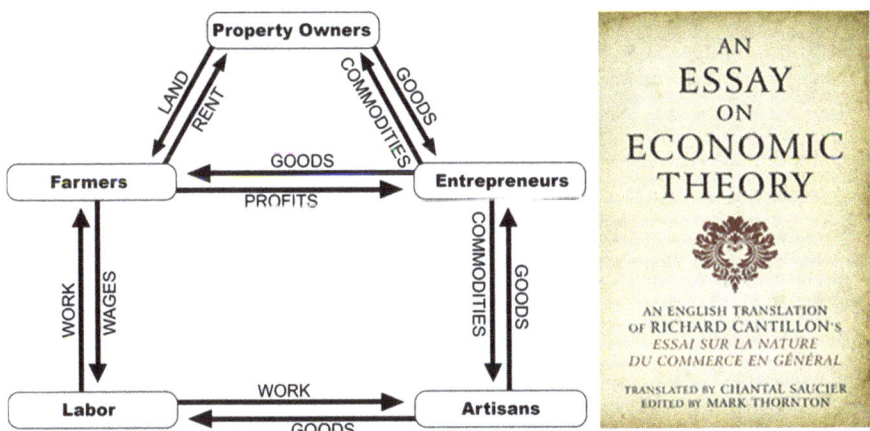

Fig. 1 James Cantillon's idea of a closed circular economy and his book

Cantillon also introduced the term "entrepreneur" as "risk-bearer," long before Jean-Baptiste Say, who gets credit for introducing this word in textbooks. For Cantillon, an entrepreneur is "a man with known expenses and unknown income." The term is used in his graph of the circular economy. Cantillon's writings were forgotten until they were rediscovered by William Stanley Jevons (who called them "The cradle of political economy") in the middle of the nineteenth century.

From a current perspective, three things are noticeable in Cantillon's work. One is the lack of any link to macro-economic concerns of nations (inflation, unemployment) or to moral concerns (poverty, usury). The second is lack of any link to external factors affecting production (climate, over-population, natural resources). His "model" is closed, like most economic models since then. The third thing to notice is also an absence: finance (and taxation). Cantillon, an international banker, did not see banks as an independent component of society with a special role. He saw banks as components of the marketplace. He realized that governments have a role and that there are costs of governance, but he probably considered customs duties, tariffs, and taxation to be much the same thing.

The so-called physiocrats in pre-revolutionary France (Quesnay, Turgot et al.) may or may not have read Cantillon's book (which was circulated privately). Like him, they had nothing to say about climate or environmental factors. Francois Quesnay was a medical doctor to the Queen and (later) the King. He was also the inventor of the first "input–output" scheme (the famous *Tableau Économique*) (Quesnay 1766). Diagrammatic representations of flows, both of goods and money, are a basic tool of economics. Quesnay's scheme was more elaborate than the simple diagram in Fig. 2, including intermediate and capital flows. He also recognized at least one environmental input, namely agricultural production. He was quite worried about the decline in fertility of French farmland, due to the high taxes on land to support the lavish court of Louis XIV in Versailles. That deprived the peasants of the surplus needed to invest in maintenance.

Turgot was a follower of Quesnay (and also a tax collector) and one of the first to worry about the causes of the creation and distribution of national wealth (Turgot 1776 [1946]). Physiocrats understandably attributed national prosperity to the agricultural productivity of the land. They advocated high prices for food, free trade, and other policies to favor agriculture. In that sense, they were the true progenitors of the resource economists and ecological economists of today.

The argument about property rights is long over, so it is not worth recapitulating here. Monarchy itself is now primarily a tourist attraction. But one

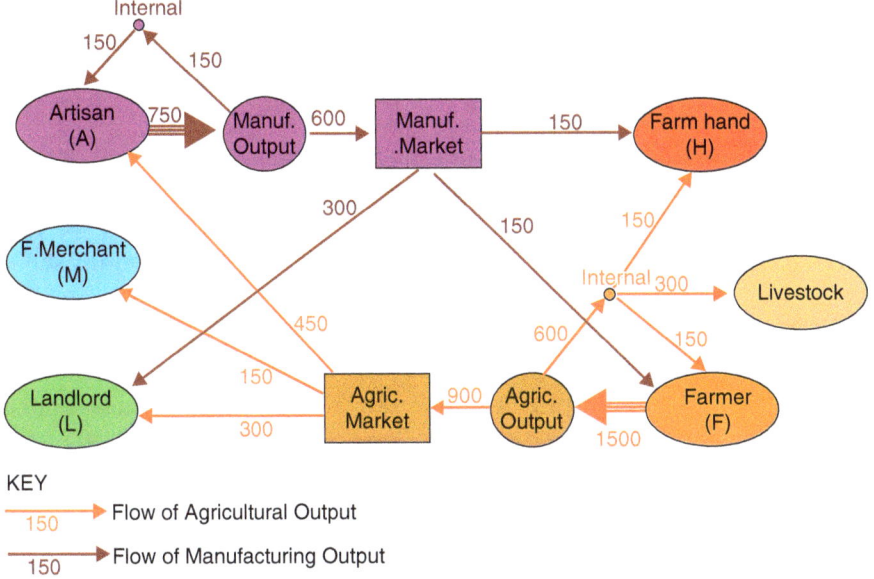

Fig. 2 Precursor of Quesnay's Tableau Economique (in English)

residue survives: The view supported by moral philosopher Adam Smith (Smith 1776 [2007]), as well as Thomas Malthus (Malthus 1798 [1946]) and David Ricardo (Ricardo 1817 [1946]). These moral philosophers also regarded land as the source of wealth, but they attributed economic wealth, in part, to manufacturing, in part to population (the labor supply) and partly to the benefits of trade.[1] But they shared one thing with Ayn Rand and all modern libertarians: the idea that the main function of government is *to protect private property*.

A short diversion on the ideas of novelist and libertarian Ayn Rand is appropriate here because her influence in right wing circles is still so great—in fact growing. Rand's primary motivation, throughout her life, was fierce defense of individualism and opposition to "collectivism" in any form. The positive aspect of her thesis is that most human wealth has been created, not by nature, but by a relatively few creative geniuses (archetypical capitalists, of course). Rand believed that invention is essentially associated with individualism, whence it cannot prosper in a "collectivist" society. There I tend to agree. But Rand interpreted any government intervention in the idealized "free market" as collectivism. In her view, pure "capitalism" is the only system geared to

---

[1] Karl Marx was the first to conceive of "capital" as an aggregate including land and animals but also including structures and machines—and money or "circulating capital."

the life of a "rational being" and the only *moral* political economic system in history (Rand 1967).

Rand's core idea, later adopted by the Libertarians, was that "*every person owns his or her life, and no person can own any other person's life.*" Her philosophical rejection of collectivism and totalitarianism morphed into a rejection of "socialism" and any social programs that involved redistribution, i.e. the use of taxes to help the less fortunate—the "moochers" (her word)—at the expense of the creative and productive individualists whom she so admired. She viewed voluntary altruism as harmless. But she saw involuntary income redistribution through the tax system as morally wrong. In her view, each person should do with his life whatever he or she wants, without any restriction, except that he or she causes no harm to others (a sentiment similar to the Hippocratic oath and to the motto "Do no Evil" of Google's founders).

Personalities apart, Ayn Rand's ideas are based on a very selective reading of the ideas of early economists, including Adam Smith ("self-interest"),[2] David Ricardo ("free trade"), J.B. Say ("laissez faire" though the phrase was used earlier), and John Stuart Mill. Mill was not much of a writer of quotable quotes. However, he wrote in his famous essay *On Liberty*: "*The sole end for which mankind are (sic) warranted, individually or collectively, in interfering with the liberty of action of any one of their number, is self-protection.*" (Mill 1869). He was an avid supporter of the union movement and even of redistribution. He also allegedly said "not all conservatives are stupid, but most stupid people are conservatives."

But J.S. Mill went on to write that we should be "*without impediment from our fellow creatures so long as what we do does not harm them, even though they should think our conduct foolish, perverse or wrong*" (Mill 1869). The problem with that statement is that the possibility of *unintentional harm* (in the form of externalities) is not considered. In fact, the very existence of harmful externalities in economics was not really acknowledged by theorists until the last years of the nineteenth century, and at that time it was regarded as a kind of minor inefficiency curable by fees or taxes.

Ayn Rand's first non-fiction book was *The Virtue of Selfishness*, a paean to unfettered pure Lockean capitalism (Rand 1964). In that book, she saw no need for social norms or a society of any kind. It was as though every person could (and should) be an independent, self-sufficient beaver hunter, oil driller,

---

[2] But not including Smith's *Theory of Moral Sentiments*.

or rancher in the wilderness. (But somehow the cavalry should always be near at hand to drive off the marauding native tribes "Indians.") Later, she edited a collection called *Capitalism, the Unknown Ideal* including a number of her own essays, along with three essays by Alan Greenspan and two by Nathaniel Branden (Rand 1967). In one place, she recapitulated J.S. Mill, again without considering the possibility of externalities and unintentional harm. In her words:

> *The proper functions of a government are: the police, to protect men from criminals, the military forces to protect men from foreign invaders and the law courts, to protect men's property and contracts from breach by force or fraud and to settle disputes among men according to objectively defined laws. (ibid, p. 47)*

In other words, Ayn Rand saw no need for a society that would offer public services such as roads, sewers, public health, education or research, never mind poverty alleviation, or environmental protection. In her view (and that of her followers, the Libertarians), all of those services are either unnecessary and should be eliminated, or should be privatized or voluntary. She has no explanation of how the public health service or the public school system could be privatized, or how it could have been created in the first place.

It is no coincidence that Rand's emphasis on the legitimacy of military force to protect people and property from foreign invaders is now applied by the USA to terrorists and jihadists located in countries far away. "Defense" has morphed into the use of force to kill "enemy combatants" and "protect American interests" around the world. Like Humpty Dumpty, we say that our words—such as "American interests"—mean whatever we say they mean.

Two pages later (p. 49) in that book, Rand says that "*the most infamous piece of legislation in American history*" is the Sherman Anti-Trust Act (1890). Why no recognition that "restraint of trade" can actually hurt people? Why the total lack of recognition that monopolies cause harm (by preventing innovation and keeping prices too high), as all economists know or should know? Why is it that, in her super-hero (John Galt's) imaginary world, the tycoons—like Rockefeller, Carnegie, and Mellon—deserve legal protection but not the little guys those tycoons ran over and squashed on their way to the pinnacles of success? You will have to read her works to get the answers.

Please forgive one more diversion concerning the issue of inequality. One religious group, the Society of Friends (Quakers), founded by George Fox in 1652, included as a fundamental moral belief that *all human beings are equal in the eyes of God, and that all deserve equal respect*. It was this belief in equality between classes and genders that led to their famous practice of addressing each other, and others, using the lower class form of address ("How does thee do today?") instead of the upper class "you" as in ("How do you do today?"). The purpose was a refusal to acknowledge distinctions of class or rank.[3] It was this egalitarianism that made Quaker merchants treat all their customers alike by refusing to bargain and charging the same fair price to every customer. Now this "one price for all" policy is almost universal in the OECD world.[4]

Nevertheless, in the eighteenth century, some Quakers did engage in the slave trade, and some owned slaves, despite those core moral beliefs in equality. During that century, there was an increasingly active debate within the Society of Friends. It culminated in a formal decision by the Philadelphia Yearly Meeting to take a collective stand against slavery. This happened in 1758. The London Yearly Meeting followed in 1761. By 1776, the Society of Friends formally prohibited slave-owning or participation in the slave trade as "Acts of Misconduct" justifying expulsion from the Society.

The Quakers were the first Abolitionists. In subsequent decades, they were also the most active participants in the "Underground Railway" to help escaping slaves from the South (in the USA) to reach sanctuary in the North. They probably deserve the most credit for ending slavery in Great Britain, which happened in 1807. Quaker factory owners like the Fry family and the Cadbury family were also pioneers in providing welfare services, including housing, for their workers. However, the Quakers also respected private property. They were not "socialists." They never advocated land redistribution, for instance.

The rise of socialism as a political philosophy may have originated with the Jacobins, in France. They, like the Quakers in England, believed that all human beings should be regarded as equally worthy of respect. They rejected rank differentiation in clothing, for instance. And before the revolution of 1789–93, they also respected private property. In retrospect, it is clear that the "winners" of the French Revolution were the bourgeoisie and the entrepreneurs (both French words, after all). But the exigencies of the "reign of terror" probably confused popular opposition to Aristocracy with opposition to the ownership of wealth, meaning land. This seems to have morphed into a contest between the bourgeoisie and the proletarians (another French word). Marx saw that as a "class struggle."

---

[3] This mode of speech later served to distinguish the Quakers from everybody else while Anglo society adopted the upper class "you" mode of address for all. In recent decades, the French have stopped using "tu" and the Germans have abandoned "du" to social inferiors, except children.

[4] The no-bargaining practice ("one price and returnable") turned out to be very good for business because it engendered trust. Quaker merchants did very well, and some (like John Wanamaker, in Philadelphia) became wealthy.

A Conflict of Ideas, Continued

The rise of "Socialism" as an organized political movement is credited to Karl Marx and Friedrich Engels, though Marx began as a Hegelian critic (of Proudhon, for example). Marx, in particular, was an economist of the first rank although his "labor theory of value" failed to allow for either the role of natural capital or the important contribution of cooperation and risk-taking. He recognized the importance of natural capital, especially coal, as source of surplus value and capital accumulation. Yet, he over-valued the role of "labor," in the sense of human muscles as this role was understood by the workers—the proletariat—who joined the movement named after him.

Marx and Engels together wrote "The Communist Manifesto" (originally Manifesto of the Communist Party). It was published, in 1848, just as the Revolutions of 1848 began to erupt, and undoubtedly helped to amplify the eruption. The Manifesto presents an analytical approach called "historical materialism" to the "class struggle" and the conflicts of capitalism and the capitalist mode of production, rather than a prediction of communism's potential for political influence. The Communist Manifesto was later recognized, nevertheless, as one of the world's most influential political documents.

Marx's political theories were also seriously faulty in certain respects. The main faults (looking back) were as follows: (1) He thought that production could, and should, be planned by the State to maximize the use value of output. (2) He thought there was no need to reward innovation and risk-taking in a planned economy because those attributes are present in every person. (3) He thought that State ownership in a democratic society would automatically assure a fair distribution of the products of the economy, each worker to be rewarded according to his work contribution and his need. Finally, (4) he thought that socialism was a transition to communism, and that it must incorporate vestiges of capitalism, thus sharpening the contradictions and triggering periodic crises. He was wrong on all of those beliefs, except the crises. He was pretty much right about degradation of nature and exploitation of workers.

In Fig. 3, the proletariat, at the bottom of the societal pyramid "works for all and feeds all" in the Marxist view, exactly the opposite of Ayn Rand's view of Atlas, the mythical giant (entrepreneurial genius) who carries the whole Earth on his shoulders and creates all the wealth.

Looking back, the political triumph of Marxism in Russia, in 1917, was almost a freak. The circumstances were that Russia was exhausted. The Tsar and his family were out. Moderate socialists (Mensheviks) took over when the Tsar resigned, but they were unable to agree on policy, even though virtually everybody wanted to stop fighting and sign a peace treaty with Germany. The difficulty for the Mensheviks was that the Western allies (Britain and France) wanted Russia to stay in the war, and there were factions in the Russian

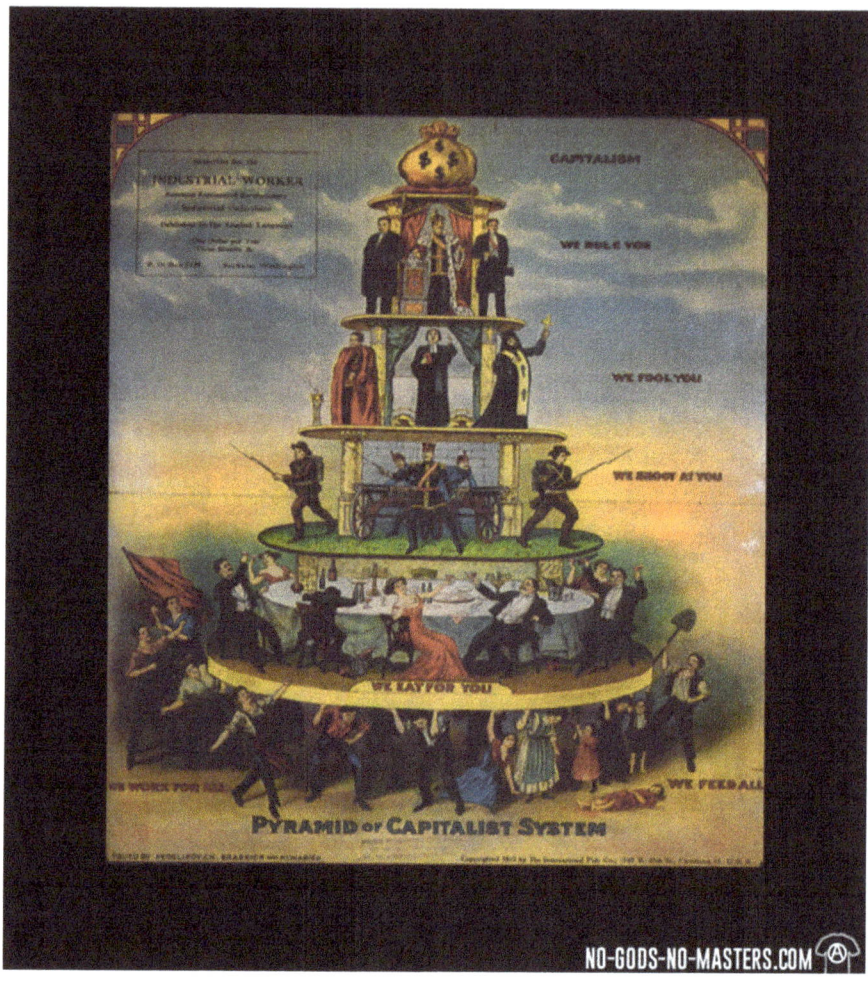

Fig. 3 From a 1911 Industrial Worker (magazine)

government wanting to maintain good relations with those countries after the war. Meanwhile the Germans also wanted Russia out of the war so as to concentrate its forces in the Western Front. Germany saw the Bolsheviks, who were refugees at the time, as the way to break the deadlock in Moscow. They facilitated Lenin's return to Russia with his inner circle, in a secret train, and the gamble paid off. The Treaty of Brest-Litovsk was signed, and Germany was free to move its armies from the Eastern Front to France. That move, in turn, probably persuaded the United States to enter the war.

The triumph of the extremist Bolsheviks over the moderate Mensheviks had other consequences. The Bolsheviks believed in central planning and state ownership of "the means of production." But when mines and factories were actually nationalized in Russia, after 1917, the owners (including Ayn Rand's Jewish family) were expelled or impoverished. For the first 20 years, the workers in those nationalized factories were probably better off than they had been under the Czar's rule. But, under Stalin's rule—with central planning—the State under-invested in the nationalized industries and failed to encourage, or reward, innovation. As time passed, they competed less and less effectively with privately owned industries elsewhere that did invest and innovate.

All three of Marx's basic assumptions about how State ownership would work turned out to be false, which is why his fourth assumption also turned out to be wrong. Thus, political socialism failed when it got power. Was this due to a flaw in the socialist idea? I think it was due to the underlying neglect by the leaders of technological change and their intellectual assumption that resources can be allocated better by planners than by free markets.

The other sort of socialism has been called "utopian" for good reason. The most effective British reformer of the early nineteenth century was Robert Owen, a Welshman who had managed cotton mills in the Manchester area. In 1799, he visited the New Lanark Mill in Scotland in 1799, fell in love with (and married) the proprietor's daughter, Carolyn Dale. He then bought the mill with partners from Manchester. When he bought it, the workers were mostly uneducated paupers from workhouses or prisons. They were paid (like workers in other mills) in tokens usable only in the mill's "truck" store where prices were set high to give the owners another source of profit. Owen changed that practice in New Lanark, and created the first of the "consumer cooperative" stores that are now widespread. He also was the first to propose the 8 h day in 1810, although it was not approved by his partners and not implemented nationally until much later.

In 1813, Owen had to buy out his partners, who wanted to continue the "standard" industrial practices on the usual grounds ("our competitors do it, so we have to do it in order to compete"). Luckily he persuaded Jeremy Bentham (whom he greatly admired) and one of the Quaker merchants, William Allen to invest, on the basis of a fixed return on their capital of £5000 each. After that he had more freedom to innovate. His innovations included education for the workers and their children, and worker housing. He gradually converted New Lanark into an industrial and social success story that attracted wider attention.

In 1813, he wrote a book, entitled "A New View of Society" (Owen 1813 [1946]). (It was followed by several other books in later years, developing his utopian themes.) In 1817, he joined the fledgling socialist party and cowrote a report to the committee on the Poor Law, in the Houses of Parliament.

His influence led to two failed utopian experiments during 1825–27 at Orbiston, Scotland, and "New Harmony," Indiana (1824). (The Orbiston failure was partly due to problems beyond his control.) But in later years, he tried again with a more carefully planned utopian enterprise called "New Moral World" which was started but sadly never actually built. It would have been a small town surrounding a factory. Later Owen helped to get the first *Truck Acts* passed by Parliament (1831 et seq.) Today Robert Owen is known and revered as the founder of the cooperative movement and of "utopian socialism."

Owen was not entirely alone. In 1844, the Society for improving the condition of the Working Class opened the first of its model housing estates (Fig. 4).

Robert Owen was not unique. In 1844, a "Society for the Improvement of the Conditions of the Laboring Class" opened its first model housing estates in England. This was neither a corporation (in the modern sense) nor a partnership, but rather an eleemosynary organization.

The most widely read book on economics in history (with the possible exception of "Wealth of Nations" by Adam Smith) was a book by a self-taught political economist named Henry George. His formal education ended at age 14. He made his living as a journalist and later became a politician. Yet, his magnum opus, entitled "Progress and Poverty," is taken very seriously by economists, and justifiably so. The book sold 3 million copies in English, in

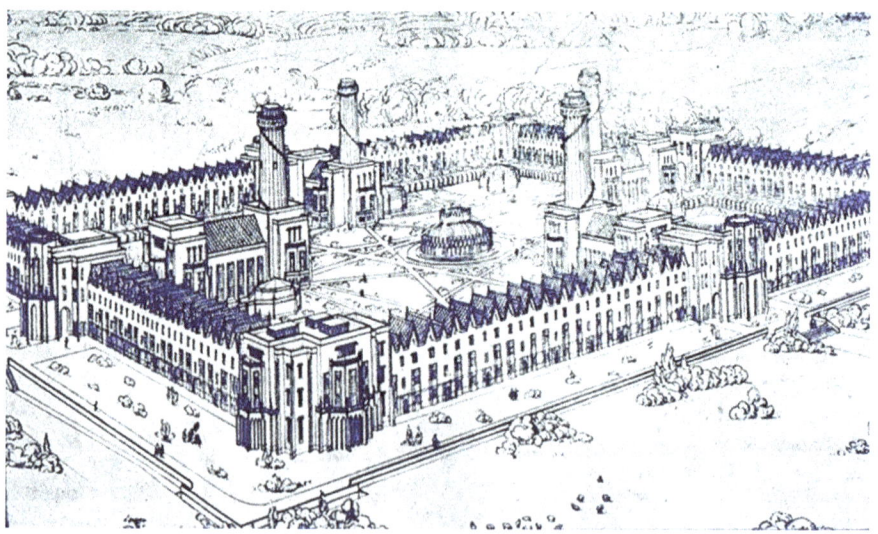

Fig. 4 Drawing of the plan for "New Moral World," by Robert Owen

the first years after publication, 6 million copies in 13 languages, by 1936, and many more since then. It is justifiably included as one of the ten all-time economics classics (George 1879). Yet, most mainstream economists, today, have never heard of Henry George.

In 1879, Henry George advocated a single tax (*impot unique*) on land values. At the time he wrote, land values were a reasonable proxy for total societal wealth. Today that is less true, and I will focus on the alternatives later. Very briefly, it argued that people should own the products of their work, but that the gifts of nature, including the fertility of the land and the mineral wealth beneath, should be shared equally among all the inhabitants of the earth or the territory.

Specifically, Henry George proposed a universal basic income (UBI), very similar to the idea that has recently been getting attention as a means of compensating for the massive unemployment that can be expected in coming years, thanks to automation and artificial intelligence. To finance this basic income, George proposed a single tax on the value of land. The following excerpt from his book (taken from the article in Wikipedia) tells it more clearly than I could. Bear in mind that he was writing at a time, after the end of the Civil War, when the United States was growing very fast. This growth was largely due to the opening up of the Western USA, thanks to railroad construction.

> *Take now ... some hard-headed business man, who has no theories, but knows how to make money. Say to him: "Here is a little village; in ten years it will be a great city—in ten years the railroad will have taken the place of the stage coach, the electric light of the candle; it will abound with all the machinery and improvements that so enormously multiply the effective power of labor. Will in ten years, interest be any higher?" He will tell you, "No!" "Will the wages of the common labor be any higher ...?" He will tell you, "No the wages of common labor will not be any higher ... " "What, then, will be higher?" "Rent, the value of land. Go, get yourself a piece of ground, and hold possession." And if, under such circumstances, you take his advice, you need do nothing more. You may sit down and smoke your pipe; you may lie around like the lazzaroni of Naples or the leperos of Mexico; you may go up in a balloon or down a hole in the ground; and without doing one stroke of work, without adding one iota of wealth to the community, in ten years you will be rich! In the new city you may have a luxurious mansion, but among its public buildings will be an almshouse.*

Henry George was making the point that rising land values in cities arise from the benefits of human activities and interactions that are attributable to high population density or high density of trade activity, including cultural activity. Yet the added value of the land created by this synergy was mostly benefiting landowners, often non-resident, who actually did nothing to add value to the land. For example, consider the Grosvenor family (Dukes of Westminster), the richest family in England, whose wealth comes from high land values in the West End of London. Or consider the Stuyvesant family of New York, descendants of Peter Stuyvesant, the last Dutch governor.

Henry George therefore proposed a tax on land values per se, stiff enough to bring in enough revenue to replace other taxes, and enough to induce landowners to sell idle land or put it to the best and most productive use. In other words, he advocated his tax on rentiers to encourage urban (and suburban) land development.

Henry George also proposed a Universal Basic Income, or UBI, to be financed by his tax on land. His proposal was not the first. That honor may belong to Thomas More in his book "Utopia" where the residents of an imaginary idyllic island shared the basic goods and services in common residences (much like life in a monastery but without the religious content). Later the Marquis Nicolas de Condorcet wrote a small book while in a French prison awaiting execution, *"Esquisse d'un tableau historique des progrès de l'esprit humain"* (Sketch for a Historical Picture of the Progress of the Human Spirit), which was published posthumously in 1795. In it he envisioned a utopian future including social insurance. It is considered one of the major texts of the Enlightenment and of historical thought.

Condorcet's ideas were simultaneously taken up and elaborated by Thomas Paine, in the last chapter of his book "*The Rights of Man*" (Paine 1791). Paine's book strongly supported the French Revolution, which is what it was mainly remembered for. But in the last chapter, Paine said that welfare for the poor is not charity, but a natural and irrevocable right. That view was radical at the time of the French Revolution (and still is in some quarters).

Charles Fourier (1772–1836) was a critic of the violence of the French Revolution, and the concept of "revolution" in practice. He became known as the founder of "*L'Ecole societaire*" or utopian socialism. He is remembered for the last of the "four apples" that conveyed wisdom and changed social history. (The first three were Eve's gift of an apple to Adam, Paris' gift of an apple to Aphrodite, and the apple that fell on Isaac Newton while he was sleeping, giving him the idea of gravity.) Fourier's apple revealed the fundamental flaws of unregulated commerce as exemplified by dumping sacks of rice into the river to create shortage and raise the market price. He borrowed some of the ideas of Robert Owen in England.

Fourier's most influential follower was Jen-Baptiste Godin, a manufacturer of cast-iron stoves located in the town of Guise, in northern France. Godin is remembered today for the concept of "Familistere," a factory that incorporated living quarters for all the workers, and was governed by democratic votes of the worker-members. Godin himself lived in one of the apartments in his complex. By 1884, his manufacturing enterprise in Guise had 2000 worker-inhabitants. Godin's ideas were very influential in "reform" circles in subsequent years. However, his enterprise failed (it is now a museum) because cast-iron stoves went out of fashion, and there was no entrepreneurial leader to carry on after Godin's death in 1888. It was a family enterprise without a viable plan for survival after the founder's death.

The Olivetti Company in Italy, manufacturer of typewriters, personal computers, and smartphones under Camilo Olivetti and his son Adriano, have been among the most advanced of all manufacturers in the world in terms of their treatment of employees, providing housing, health services, pensions, and even entertainment. Sadly, Adriano died young and since his death, the company has fallen into financial difficulties resulting from unwise acquisitions. It is no longer an independent family-owned firm, now being a subsidiary of (George 1879) Telefonica Italia.

The Rise of Corporate Capitalism and Its Champions

Profits are undoubtedly the main driver of economic growth, but organizational forms play a huge, but seldom recognized role. The failures of Owen and Godin were partly due to inappropriate organizational structures, depending too much on the leadership of a single individual. Apart from profits, in successful organizations a lot depends on collaboration between people with different skills. Even more depends upon how responsibilities are passed on from generation to generation.

There are three organizational forms that have survived over time. The first and oldest is the family business, such as a farm or small manufacturing enterprise based on land-ownership or heritable skill. Heritable skills included masonry, metal-working, shoe-making, paper-making, or dye-making, that can be passed on from father to sons (rarely daughters). Family businesses, apart from small farms, are becoming a rarity because inheritance by bloodline is a poor way of choosing leaders, and because siblings often compete but rarely collaborate by mastering all the necessary skills. Hence family business cannot compete in the long run against collaborative enterprises.

The next form of organization to emerge was the legal partnership, with an indefinite lifetime and the ability to add new members, while retaining a common purpose and a leadership selection mechanism. At any partnership, risks and profits of an enterprise are shared proportionately among the partners. Each partner is personally responsible for his or her share of financial risks. Finally came the modern corporation, discussed later although its form has also evolved.

The model partnership is probably Lloyd's of London—which began as a London coffee shop in 1686—is a kind of marriage of interests. It began as an insurance marketplace originally for bringing together shipowners and investors to form syndicates, each of which specialized in a particular kind of business. (At Lloyds, investors are called "names.") Lloyds moved to Lombard Street in The City of London in 1691, retaining its function as a market for shipping, insurance underwriters, and shipowners. For the next century, it played a major role in the slave trade. Lloyds main function as an organization was, and still is, to provide a central source of information, both to investors and to client entrepreneurs.

Consumer cooperatives are partnerships of another kind, mainly to benefit the purchasers of consumer goods by central purchasing and non-profit distribution. In this case, all partners are simply members of a club, usually with equal rights or rights based on capital investment. The first cooperative shop seems to have been initiated by Robert Owen, the "socialist" pioneer, for the workers in his New Lanark cotton mill. That was around 1810. However, the idea spread, starting in Scotland and northern England. It was formalized in 1844 by the Rochdale Society of Equitable Pioneers, each of whom contributed one pound of capital for the venture. That was a significant investment for a laborer or a small shopkeeper in those days.

Apart from consumer cooperatives, most law firms, doctors' group practices, real-estate firms, and accounting firms today are legal partnerships. In most cases, new partners need to invest capital in the firm before sharing in the profits. A partnership's earnings after costs (including salaries) are then distributed among the partners, according to some agreed rule. The rule is usually not equal, being usually determined by the founders and based on seniority, but it is not extremely unequal. (Some junior partners might disagree: the legendary allocation of income is said—by some juniors—to be the inverse of the allocation of work to the partners.)

Until the second half of the twentieth century, investment banks were partnerships. Because of their role in the financial crash in 1929 and the following Great Depression, they were legally prevented from operating as commercial banks by the Banking Act (Glass–Steagall) law of 1933.[1] That law prohibited investment banks from taking deposits and prohibited commercial banks from dealing in, or underwriting, non-government securities. Unfortunately a series of exceptions and modifications starting in the Reagan years have put an end to that separation. Today, commercial banks both take deposits and gamble with depositors' money, while their executives simultaneously enjoy the protection of limited liability, meaning that when their gambles turn out

---

[1] The Act is named for Senator Carter Glass and Representative Henry Steagall. It was arguably the most important legislation introduced by newly elected President Franklin D. Roosevelt in 1933.

badly, they do not take personal responsibility. I have more to say about this later.

The other type of company is the corporation. The word comes from Latin where it meant "a body of people," bound by some common purpose, that can survive longer than its individual members. Corporations originally required a royal charter specifying its purposes and its limits, or some modern equivalent. Churches in medieval Europe were corporations, as were municipalities (e.g. the City of London). The right to incorporate was granted by the emperor (starting with Justinian, in Rome). The rights of a corporation included ownership of property, the right to make contracts, and to sue for breach of contract. In Roman times, there were corporations for burial clubs, political groups, and guilds. The oldest known corporation in Europe, created for commercial purposes, was the copper-mining company Stora Kopparberg, of Falun, in Sweden. It was incorporated by royal charter (from King Magnus Eriksson) in 1347.

A corporation is a "legal person" with indefinite lifetime, but (since 1855) with limited liability for the managers or executives, in case things go wrong.[2] Managers may or may not be owners; if not, they are regarded as "agents" for the owners. The Joint Stock Company Act of 1844 legalized what had been going on since the seventeenth century, and allowed (for the first time) for a joint stock company to be represented in litigation by a single representative. It is widely (but falsely) assumed today that the owners of a joint stock company are the shareholders. This is not true in law, or in fact. In law, a corporation is a *legal person that (who) owns itself*, just as individuals—since the end of slavery—"own themselves."

Shareholders have certain legal rights, but they lack other rights. The rights they own may include the right to elect the Board of Directors but not necessarily with equal voting rights for all shareholders. They may enjoy the right to receive dividends, as and in the amount designated by the board of directors. Shareholder may also share in the distribution of assets, in case of bankruptcy, but only after the payment of taxes, wages, bills of suppliers, and bondholders. (Shareholders in bankruptcy cases are last in line.) Executives are "agents" who are hired by the Board to run the company. They have responsibilities as specified in their employment contracts, but no personal liability for losses except in the case of criminal behavior.

---

[2] The first limited liability law for a joint stock corporations was enacted in the State of New York in 1811. The Limited Liability Act was enacted by the English Parliament in 1855. The distinction between private and corporate liabilities was confirmed (in England) by the landmark case Academia.edu.

The fact that corporations do not own themselves does not make them legally equivalent to humans. The difference is that humans are self-aware, with individual aims and aspirations. Corporations have no self and no selfish dreams or aspirations. They are much more like puppets that dance to the strings pulled by a puppet master. However, the self-ownership of corporations can effectively disenfranchise both the shareholders and the employees. Public companies can be treated as "pots-of-gold" by a dominant shareholder who is able to control the Board of Directors and control, or become, the Chief Executive Officer (CEO). Quite a few Wall Street legends have achieved billionaire status this way. I forbear to name names. I think this is a weakness in the existing capitalist system.

The limited liability law of 1855 (UK) has one other enormous, and dangerous, consequence that bears a major responsibility for the increasing income inequality in recent history. It means that corporate executives are legally entitled to take more risks with the firm's money than members of a partnership, without necessarily suffering the consequences of misjudgments or mistakes. This distinction has been very important in the history of finance. It is also the key (for some executives) to accumulate enormous unearned wealth.

Corporations may be born from other corporations (e.g. as spinoffs), but only at the will of the executives and directors of the parent. They do not create themselves. They are usually created by founding entrepreneurs who specify their purposes and write mission statements, equivalent to charters. The founders also write the rules governing election to the Board of Directors and decision-making by the Board. Founders (and often their heirs) can, and frequently do, create ways of controlling their Boards of Directors for multiple generations, imitating family firms. Ford Motor Co. is still controlled by the Ford family. Cargill, Hershey, and Mars are a few of many other examples of "family" firms. Many, if not most, of the major German companies are still family controlled. This has created a new hereditary upper class, without formal titles or other emblems of rank.

A common feature of the life history of financial tycoons is good luck, self-confidence to an extreme degree, detailed knowledge of some business, and clever use of financial "leverage," especially the latter. You may wonder: "what is leverage and where can I find it?" As for the first question, leverage is the ratio between the amount of money available for investment (or speculation) vs. the capital reserve. The leverage for US banks is currently around 10–1, though in 2008 it was much higher. As for how you can get it, there are some gimmicks discussed later, if you are rich enough. Or start a bank of your own.

Reverting to the life history of a modern Horatio Alger tycoon: our boy-hero "M" has found a market opportunity. He sells something tasty and edible (call it X) from a bicycle. At first he is very cautious. He successfully avoids serious damage by competitors. He manages, after some searching, to find a "sweet spot" for selling more X than a single person (or two) can provide from a single bicycle. In the course of some months, he saves enough money to buy another bicycle or to partner with a person who has a bicycle. In a year they find a way to make the product more exciting. They agree to combine efforts with another X-seller, to cover more territory.

The next 5 years or so is characterized by slow but steady growth by accretion and capital accumulation. M improves X and adds Y to his menu. He experiments with combinations, like XY, XYX, YXX, and X + Y. He hires an assistant to manage the Y side. Then M adds Z to the menu of products and there are more combinations. He needs more space. He needs more logistics. He needs more assistants. One of them becomes his principal assistant. He needs an accountant. He needs a tax advisor. He needs a lawyer. The business is now a success. It is registered in some State.

If M is like most micro-entrepreneurs, he takes most of the profits of his business as personal income, meets a girl, gets married, starts a family, buys a nice house with a pool and a Mercedes-Benz or a BMW. If M is more ambitious, with fire in his belly, he invests those first profits in expanding the business. The business pays as little tax as possible on its profits. M's business needs working capital (cash) to operate. Employees and suppliers expect to be paid on time, but customers do not always pay on time. He borrows money from a local investor/moneylender (a relative? local bank?) using "receivables" (money he is owed by customers who have not yet paid) as collateral. This is very tricky because moneylenders will not accept goods or real estate (except in rare cases) as collateral, however salable. But, after another 5 years of modest but steady growth, the local bank gives him a line of credit, based on that 5-year history. If the XYZ business continues to grow, the line of credit will expand in proportion.

Then comes the "moment of truth." The XYZ business is doing well, and the future prospects are good, but it is still a personal venture. His customers know him personally and trust him, his employees are starting to bet their futures on him. Can he (or she) depersonalize the business and run it from a distance? If so, it needs middle managers. It needs procedures and rules. It needs professionalism.

It also needs major capital investment to grow faster. The next step for M is to incorporate and issue shares to the public, through a broker. Nowadays this is complicated and requires a lot of expertise. Or he can approach a venture capital firm. Let us assume that he finds a backer who pays cash for equity shares in his enterprise. If he is very smart, he can keep most of the voting shares for himself and his inner circle. At this stage, he tries to attach a high value to his company, based on a business plan and some new concept of wealth creation. The shares are marketable, and XYZ Corp now has a financial identity, a market value, and record of growth.

Now suppose M sees an opportunity to buy shares in ABC Corp, a firm in a similar but synergistic line of business that is "undervalued" because it is not growing as fast as it used to, or could under different management. Its leadership is getting old and has run out of ideas. Now M uses tradable XYZ shares—which have a market value—as collateral for a bank loan for the ABC company. M (who is now the CEO) uses the proceeds of the loan to buy a small fraction (say 5%) of the shares of ABC, a much larger company. He buys, if possible, from the largest shareholder, who is getting old and ready to retire. This way, there is no larger shareholder to oppose his demand for a directorship.

After a period of time to learn the strengths and weaknesses of the ABC company, M—now a director of ABC—proposes a merger with XYZ, focused on fixing those weaknesses. He proposes to use the profits of ABC to buy back the shares of its stock. Or, if the interest rate is low, as now, ABC will borrow the money to purchase its stock, while distributing cash to its shareholders, as dividends. This will increase his (XYZ's) relative share of ABC, and it will increase the ABC company's profits per share. The market price of the ABC shares will (probably) rise.

At this point either XYZ can merge itself with ABC, by exchanging XYZ shares for ABC shares at a favorable rate. Or it will sell its 5% investment, taking a cash profit (thanks to the ABC price rise) that adds to its own apparent earnings. The market will typically "reward" this earnings increase by increasing the market price of XYZ stock. Now M has more collateral available to borrow with, and he can go looking for another opportunity to buy into another larger company.

If his "takeover" (by merger) of ABC is successful, he will use the assets of ABC to borrow money from the bank to buy a small fraction of the shares of a still bigger company (DEF), and the scenario will be repeated. In a few more years, M, who is very good at his game, can end up in control of a very large corporation. In that case, he can use the assets of that company to invest in assets that will enrich him personally (as Victor Posner used the assets of Sharon Steel Corp. to enrich himself and bankrupt the company). Or he may force that large corporation to sell itself to a still larger one (as H. Boone Pickens forced Gulf Oil Co. sell itself to Chevron). In every case, M uses the assets of the "target" company as collateral for bank loans to buy more shares of still larger companies.

This capsule history applies quite well to some of the "activist" investors of today. It illustrates one of several means of obtaining a high degree of leverage for a small original investment. But if, for reasons beyond his control, the value of the collateral for the original loan declines, it can start a chain reaction that has widespread negative consequences.

There are other ways to get very rich that have slightly less negative social consequences. Consider Warren Buffet, the most famous "value investor," now 87 years old, who made $76.7 billion, starting from nothing. Well, not literally nothing, but from a middle-class background in middle-America (Omaha, Nebraska). What he did, early in his career, was to raise enough money to buy a failing shirt company with a reputation for quality (Berkshire-Hathaway) and use it as a vehicle to buy other companies. His two long-term rules, based on his own statements, have been (1) avoid competition and (2) minimize investment in the real economy ("bricks and mortar") (Harding 2017). He also advocates holding his investments for the long-term, not looking for short-term profits.

Buffet's first rule has been set forth semi-poetically (still in his own words) *"I want a business with a moat around it and a very valuable castle in the middle"* (ibid). Buffet's instructions to his managers are *"to widen the moat."* That rule was borrowed directly from J.P. Morgan, who promoted trusts (including US Steel) precisely to avoid "destructive competition," e.g. Seavoy 2013, p. 250. A best-selling book called "Blue Ocean Strategy" has essentially the same message, presented in very positive way (Kim and Mauborgne 2004). The message of that book is spelled out clearly in the subtitle: *"How to create uncontested market space and make the competition irrelevant."* The way to create uncontested market space can obviously be described as *"digging a moat"* or finding a patch of *"blue ocean"* where there are no competitors—or (harking back to the epic of Gilgamesh) *"building a wall."*

Economists have a name for the result of elimination of competition, viz. *oligopoly* or *monopoly*. After all, that is what patents and copyrights are supposed to do, at least temporarily. Thomas Alva Edison was the world champion in this field: he accumulated 2332 patents, worldwide, of which 1093 were in the USA. Unfortunately, the word "monopoly" has had bad vibes ever since the Sherman Anti-Trust Act of 1890, which prohibits "restraint of competition." Of course, many of the inventions that drove the industrial transformations described earlier in this book were originally patented so as to restrain imitation and competition. Indeed, patent disputes occupy an important place in industrial history (and still do.)

But modern monopolies or oligopolies are not always based on patents. Many are based on know-how and expertise. Coca Cola is the prime example.

But Microsoft, Amazon, and Google owe their corporate dominance to software (computer operating systems, browsers, "search engines," and "cloud" software) protected by thousands of patents, some owned, some licensed. Visible monopolies, such as public utilities, telecom companies, and "platform companies" (e.g. Microsoft and Google) are widely hated, despite the essential services they provide.

Evidently there are two sides to the story: some services seem to be natural monopolies (like electric power generation, telecommunication, public health, public safety, and national defense), while others are criticized precisely because they discourage diversity and innovation. Society clearly does not want a variety of electric power producers, some providing DC, others providing AC with different voltages and frequencies. Similarly, society wants health standards, standard weight and distance measures, and standard exchange rates. Standardization and centralization are clearly preferable. But should we all be forced to use Microsoft *Word* and Microsoft *Outlook*?

However, extreme twentieth century political capitalism (as opposed to historical and theoretical capitalism) began in the USA as a reaction to political socialism. As an antidote to "socialism"—equated to opposition to private ownership of land and "the means of production"—it emphasizes private ownership of everything possible, free markets with no regulation (except law enforcement), low taxes, and small government with minimal bureaucracy. It assumes (contrary to evidence) that the wealthy people are the "job creators" whence the government should help the rich to get even richer, so that the middle class and the poor can gain from "trickle down." This is essentially the platform of the Republican Party in the USA, at least as it was before the arrival of Donald Trump.

Unfortunately, "trickle down"—as it worked in the days of Vanderbilts and Rockefellers—is very inefficient as a way of helping the people at the bottom of the pyramid. This is because most ultra-rich people, then and now, live modestly and hide their wealth. They do not spend much of their money on things produced in factories (except the occasional large yacht or private jet). They spend their time running their businesses from a distance, always through chains of intermediaries, using the tools of financial engineering: The game for most is maximizing profits, in the narrowest—purely monetary—sense, and cutting overhead costs to the bone. Cutting costs, in the real world, always means reducing employment and moving all activities that can be moved, to low-wage states or countries. Corporate profits today are therefore at an all-time maximum, while wages in the USA have stagnated for more than 30 years.

The Rise of Corporate Capitalism and Its Champions

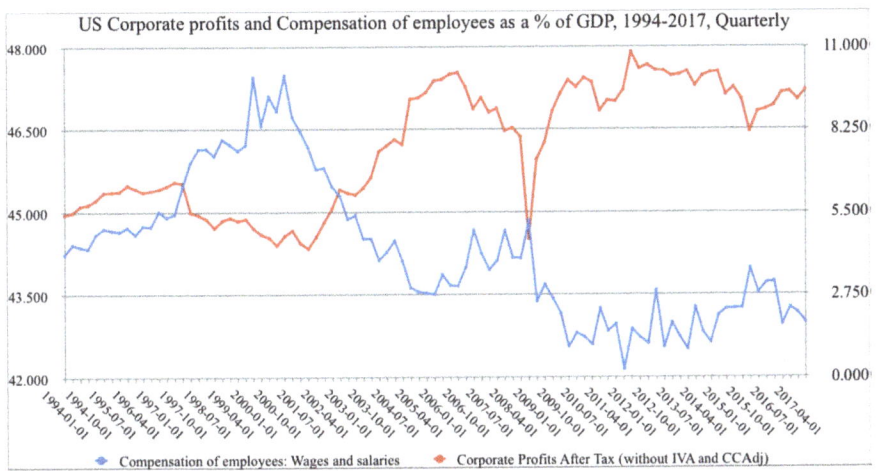

Fig. 1 Corporate profits and employee compensation

In fact, in 2017 US corporate profits exceeded 9% of GDP, for the first time. Thanks to the assumed future cut in corporate tax rates (from a top rate of 35% to the new "competitive" rate of 20% or possibly 22%), US government revenues will fall significantly in the next 2 years, while already record profits will increase still further—according to the experts—from 9% of GDP to 10% or more.

There is a close (inverse) historical correlation between corporate profits and employee compensation (Fig. 1). When profits fell, as a fraction of GDP (as happened from 1997 until 2000), employee compensation—wages and salaries—rose. It works like a children's teeter-totter. The same thing happened from mid-2006 through mid-2008. Thereafter, recovering profits were accompanied by further declines in wages and salaries, starting in 2009. Overall, since 1994, employee compensation (salaries and wages) in the USA have shrunk from 47% of the GDP to 43%. During that same period, corporate profits, after tax, have doubled, from 4.5% of GDP to 9% of GDP. It sure looks as though the increased profits are coming straight from the pockets of the employees and workers.

Can that apparent "teeter-totter" linkage be real? Well, in business there is an obvious link between profits and costs—which means wages (plus debt service and energy). It is conventional wisdom that "raiders" and "activist" investors (like Nelson Peltz and Carl Icahn) always seek to increase profits. They do so by finding ways to cut costs. Cost cutting is usually explained in terms of focusing on "core competence," selling land or buildings, eliminating unprofitable branches or unprofitable products, and (yes) laying off older workers who are approaching pension age.

Other devices in the cost-cutting bag of tricks include cutting employee health benefits, reducing R&D, reducing employee training, converting pensions from "defined benefit" to "defined contribution," and more of the same. The bottom line is that in virtually all cases of "corporate restructuring," the employee head-count is reduced. Cost cutting means cutting wages and salaries (and jobs), which are the major costs of business in most sectors of the economy. They are also, as Henry Ford knew, the source of demand for products and services.

Now we come to jobs and economic growth. The Gross national product (GDP) is defined as the sum total of all payments for goods or services in the economy. It is very helpful to think of the GDP as a "pie." The pie is traditionally divided (by economists) into two parts. The first, and largest, is called "returns to labor." This includes all wages and salaries paid directly to workers, plus the sum total of "entitlements" (from government), such as pensions, social security, health benefits, educational benefits (such as the GI Bill and "Head-Start"), and unemployment benefits.

The other (smaller) part is called "returns to capital." It includes interest payments on loans (debt service), dividends, rents, and royalties. During long periods in the twentieth century, returns to labor accounted for 70–75% of GDP, while returns to capital fluctuated around 25–30% of GDP. In fact, that relationship has long been regarded by economists as a "stylized fact," not really like a constant of nature, but quite reliable. Since the 1980s, labor's share has been in decline

However, since the 1990's that comfortable and relatively constant relationship has changed, partly due to regulatory and other changes in the financial system, especially monetary policy. Only the central bank—the Federal Reserve Bank (FRB)—has the power to change the returns on (costs of) capital. The FRB does this by raising or lowering the "Fed Funds rate," which determines the so-called prime rate on which all loans are based (Fig. 2).

Fig. 2 The history of US interest rates since 1930

The history of US interest rates is quite interesting. But for now only the most recent years are relevant. Note that the all-time peak occurred in December 1980, when the FRB under Chairman Paul Volker decided *"kill inflationary expectations"*—which had been rising steadily since the 1960s—*"once and for all."* It seems to have worked because US (and global) interest rates have declined since then, apart from some short-term fluctuations.

Speaking of fluctuations, it is worth noticing that in the years 2005–06 Alan Greenspan's Fed raised the Fed Funds rate, hence the prime rate, from 4% p.a. to 8% p.a. in several steps. This increase sharply increased the cost of borrowing. The Fed made that move just before the financial collapse in 2008, which forced another sharp reduction in rates. There are economists who argue the ill-timed rise in rates actually started the slowdown in house purchasing that led to increasing defaults and kicked off the 2008 crash. Since 2010, the prime rate has been kept at 3.5% until the first increases in 2016. But the point of this chart is that the cost of capital—and of home mortgages—took a sudden leap in 2006–07 just before the crash of 2008. But now (November 2019) interest rates are declining again.

But what can be said for sure is that when the prime rate rose in 2006 and 2007, so did the overall returns on capital during those years. That theoretically left less of the GDP pie (wages and salaries) for the workers. Yet workers compensation also increased slightly during those 3 years. This also happened because of the sub-prime real-estate construction bubble. (Construction workers get higher than average wages.) That led to a major imbalance in the financial system: simply, *rising interest rates cannot co-exist for long with rising wages*. The bubble could not last, so it did not. Wages fell and unemployment rose dramatically in the following year, 2009.

Yes, it is true that civilian unemployment has been declining in recent years. It is now (2020) even less than 4% per annum. That is regarded by academic economists as very good news. What the unemployment statistics do not show, however, is that when unionized middle-aged construction workers, or auto workers, are laid off, the alternative jobs they can get, like security guards or truck loader/unloaders (sometimes called "McJobs"), pay a lot less than their old jobs and often cannot support a family. *So unemployment can decline, and profits can rise (as they have) without any significant increase in wages and salaries.*

There have been brief periods of increasing egalitarianism in the past, mainly after wars and great social upheavals. In contrast, the US history, since 1979, is of increasing inequality as shown in Fig. 3. What this graph shows, if you look carefully, is that from 1979 through 1985 almost everybody was a little better off, especially the bottom half. But since that time, over 60% of

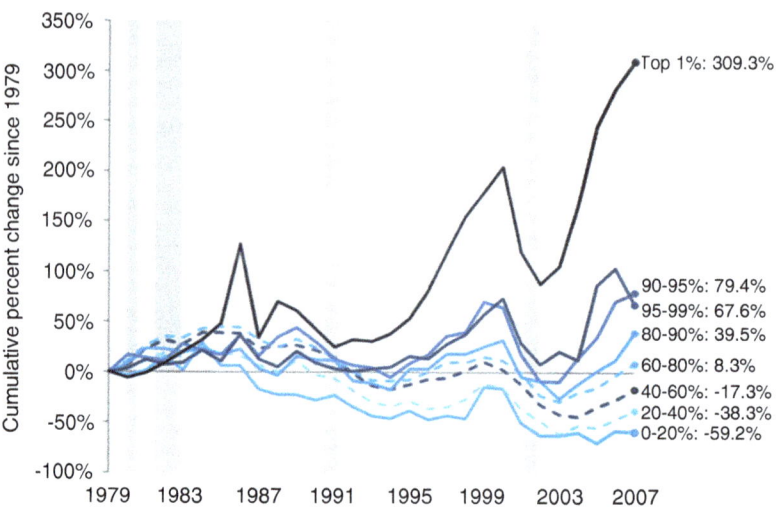

Fig. 3 Cumulative change in real annual household capital income, by income group

the households in the USA were worse off in 2007 than they had been in 1979. The bottom 20% of the households were (on average) 59% poorer in 2007 than they had been 28 years earlier, whereas the top 1% were more than three times richer. I have not been able to find an update of this graph, but other evidence suggests that the pattern of increasing inequality has continued.

A Brief History of Financial Booms and Bubbles

Capital losses can occur for a variety of reasons. One common example in history arises when a king overspends (usually on wars) and refuses to pay his creditors. Kings were able to do this because if a King wanted to borrow money, it was dangerous for a banker or a Jew to say "no." This resulted in bank failures—possibly in other countries—that hurt a lot of other banks and their depositors, the merchants and traders. Edward II of England did that with disastrous effects on Florentine banks in the year 1345. Later, the Spanish and French Kings—especially Philip II and Philip III of Spain and Louis IV of France—did it again and again in the sixteenth and seventeenth centuries (Reinhart & Rogoff 2009). The wars of Spain destroyed the formerly powerful Medici, Fugger, and Welser banks in the process. (The Rothschild family avoided that fate in 1812–15, partly by foreseeing the dangers and hedging their loans to likely royal losers, and profiting from the collapse of Napoleon's over-stretched empire itself) (Morton 1962).

There are two kinds of bubbles. All are based on crowd psychology, but one applies exclusively to investments in ephemera. There are several interesting historical examples, starting with tulips and South Sea trading monopolies. The current example (in my view) is "bitcoins" and the other so-called crypto-currencies today. The other kind of bubble is over-investment in real enterprises, such as canals, railways, Florida land, or unprofitable "Dot.Com" companies.

A crucial feature of economic history is the frequency of financial crises. This peculiar history has been extensively documented in Charles Mackay's book "Extraordinary popular delusions and the madness of Crowds" (Mackay 1841 [1996]). In 1636–37 in Holland, there was a famous craze for investing in tulip bulbs. Two breeds of tulips were especially prized: one was called *Semper Augustus* and the other was called *Viceroy*. Tulips had been a gift from the Sultan of Turkey to the Emperor Charles V in 1554. But they grew well in the Netherlands and had become the fourth most important export of the country. Between May 1636 and February 1637, the price of bulbs for those particular tulips rose spectacularly as almost everybody in the country seems to have invested in them (Mackay 1841 [1996]).

The bubble then collapsed, leaving some speculators richer and many others poorer. Figure 1 shows *Flora*, the goddess of flowers, with flag, her arms full of tulips, followed by Haarlem weavers who have abandoned their looms, riding gaily to their inevitable destruction in the sea. The bubble-and-bust phenomenon illustrated in this painting repeats itself over and over in economic history.

Fig. 1 "Wagon of Fools" by Hendrik Gerritsz Pot (1637)

Early in the eighteenth century, John Law, a Scotsman, had the bright idea of creating a bank whose capital consisted of bonds of a company (existing only on paper), expecting large future profits from a trade monopoly to be granted by the Sovereign, or by Parliament. This idea was proposed at first for Scotland and was rejected. But it was later accepted in France where the government was insolvent due to profligate spending by the "Sun King," Louis XIV. There it became the basis for the Banque Générale and the Mississippi Company (later "Mississippi Bubble"). From 1718 to 1720, the Banque Générale sold shares of the Mississippi Company to the public. The prices rose spectacularly, thanks to greatly exaggerated and ephemeral profit expectations.

Those expectations were based on the large (but mostly uninhabited) areas then controlled by France in North America. The French territories considerably exceeded, in area, the British colonies confined along the Atlantic coast (see the map; Fig. 2). The Mississippi Company was not a silly idea, if you

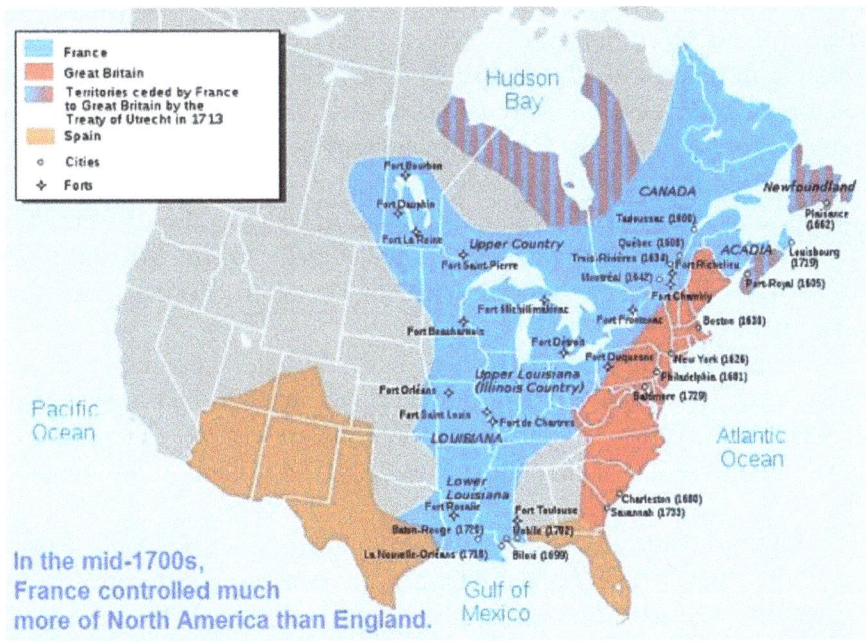

Fig. 2 Mid-eighteenth century European claims in North America

have a very long time horizon. But the bubble collapsed when the number and value of the shares issued far exceeded the unrealistic short-term revenue expectations that had been fed by rumors. John Law was (perhaps unfairly) discredited by the bubble's collapse. He had to flee from France.

Nearly a century later, Napoleon Bonaparte eventually sold the French territorial claims in North America to the United States during Jefferson's presidency, in 1804. It was still mostly unexplored land (except for the region around New Orleans) exporting beaver skins for hats and a few other forest products.

But the idea of promoting joint stock companies based on the expectation of future trading profits had also already crossed the English Channel. A similar bubble, the South Sea Company, also based on a private trade monopoly. The "South Sea bubble" started in 1711 and the share prices rose sky-high, briefly, until the inevitable collapse (see Fig. 3). Again, the shares were sold to the public based on unrealistic expectations of short-term monopoly profits from trade with Chile, Peru, and the "antipodes." The profits were supposed to pay off the English public debt held by the Bank of England.

In fact, the "dividends" to early investors were paid from the money received from later investors. Those shares were (briefly) treated as currency. After the collapse of this "Ponzi" scheme[1] (1720), the privately owned Bank of England (the official banker and gold depository for the British Sovereign at the time) bought the South Sea bonds for silver. (It is alleged that bank partners and employees of the South Sea Company who were paid for their shares in silver, at the front of the bank, had to carry it physically to the back of the bank to keep the flow going, until the panic subsided.) This coup also made the BoE the official lender to the parliamentary government, for the first time. (The BoE did not become the official lender of last resort, during crises, until much later.)

The South Sea Company was rescued by the BoE. It did not disappear until much later. Its original building burned down in 1826, with most of the records. But the company continued in two lines of business, the burgeoning slave trade and arctic whaling. This continued to be profitable until sometime after 1838.

Another consequence of the South Sea bubble was an Act of Parliament, the "Bubble Act," which explicitly limited the number of investors in a new company to five individuals. (This law made financing of new enterprises

---

[1] Charles Ponzi was a Boston-based "developer" who sold building lots (23 to the acre) in a largely fictitious real estate development "near St. Augustine" in Florida—but actually 65 miles west of the city— during the Florida land boom of 1919–1920 (Galbraith 1954, pp. 9–10).

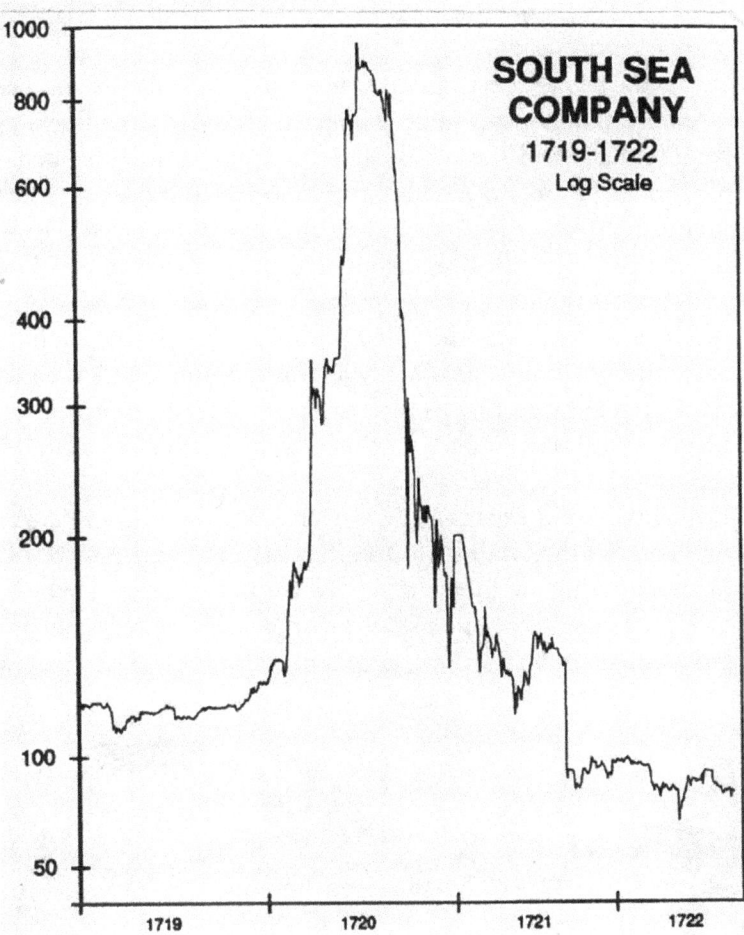

Fig. 3 Price history of the "South Sea Company" shares

difficult and undoubtedly acted as a restraint on economic growth. The Act was repealed a 100 years later, on the eve of the railway expansion, in 1825.)

Distance and poor communications created opportunities for fraud. The most spectacular example was a large "sovereign" bond issue by the imaginary country of "Poyais" that was sold to wealthy (but ignorant) English investors in 1822 by a Scottish adventurer, Gregor MacGregor, who had fought for the independence of Guatemala before going into business for himself (Morgan & Narron 2015). The bond revenues did not go to Guatemala, but to

MacGregor. This was only the most interesting of the many financial frauds perpetrated at the time.

Eventually the South American investment bubble burst (in summer 1825) as Latin American bonds lost value as collateral for loans, the English issuing banks became shaky and—as usual when this kind of thing happens—there was a panic as depositors demanded their money. The damage spread and, before it ended, 10% of the banks in England and Wales failed (Fig. 4).

During the panic, the BoE actually raised the discount rate in December 1825 to protect its profits—it was a private bank—and the value of its gold. This tightening of credit was the worst possible policy, in the circumstances, and it made the situation worse.[2] Only after some large London banks failed did the BoE step in by cutting its discount rate for loans to provincial banks. (The "discount rate" is the interest rate paid on loans from the central bank to other banks.)

To finance the rapid expansion of cotton planting in Georgia, Alabama, and North and South Carolina in America, driven by the growth of cotton textile manufacturing in England, many cotton planters borrowed from more than one lender. The lenders, at the time, had no means of comparing notes. In effect there was an uncontrolled increase of credit that was not

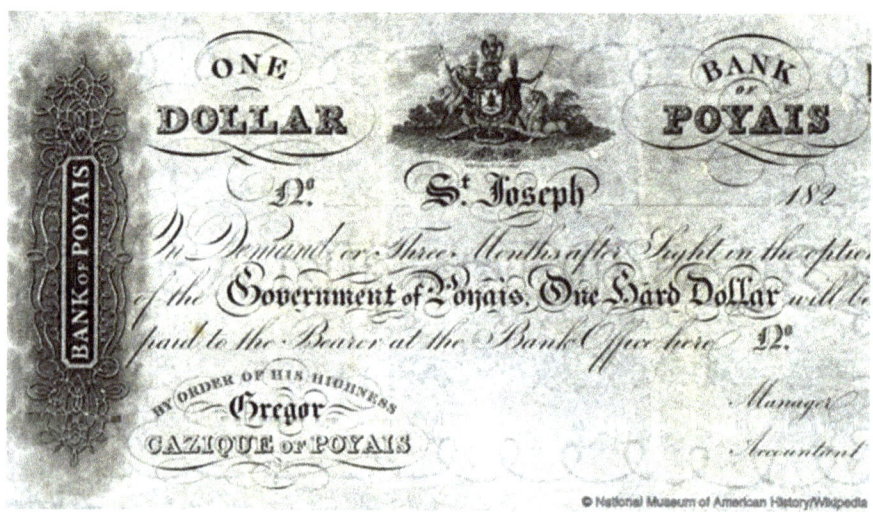

Fig. 4 A "dollar" bill issued by an imaginary country

---

[2] The US Treasury made the same mistake in 1931, in order to curb the loss of gold. This move converted a short-term recession into the Great Depression.

underpinned by real collateral. When a borrower could not pay, the lender was at risk. When a default occurred (as always happens after too much speculative investment), the result was a massive run on the banks, not only in London but—thanks to the "domino effect"—in the Netherlands and Scandinavia.

The worst example was the failure of the Scottish bank of Neal, James, Fordyce, and Brown, in 1772. It was in trouble because of losses due to speculative "shorts" on the stock of the East India Company. The bank was using depositor's money. Alex Fordyce, one of the partners, defaulted (and fled) to avoid payment. The consequence, thanks to contagion, was an immediate run on all the banks in London and Amsterdam. That led to the failure of 20 English banks (Narron & Skeie 2014). The panic was only ended by a massive liquidity intervention by the Bank of England. This happened fortuitously even though the BoE had no legal requirement at the time to act as lender of last resort (LLR).

There were other noteworthy bubbles, e.g. in canal shares (1793–95), the South American investment bubble (1822–25), and the "railroad mania." The latter had a brief peak in 1836 followed by a much bigger one in 1845–46, followed by a massive "crash" of railroad share prices that hit bottom in 1850. In every one of those cases, the collapse of the bubble left many investors poorer and—in several cases—nothing of permanent value to the economy as a whole. The canal investments of 1793–1795 and the railway investments of 1835–1850, however, were different. They left a residue of value (the physical canals and rails) that continued to support the British economy for decades.

There have been a number of important financial bubbles in the United States that also qualify as ephemeral, in the sense that nothing of permanent value was created, but enormous losses followed. The panics of 1857, 1884, and 1893 were all directly caused, or made worse, by gold shortages, foolish legislation in favor of silver miners, and lack of financial regulation. Over-investment in railroads was the primary cause of the panic of 1873 and again in 1884. Had there been a US central bank at the time, and a currency based on "faith and credit" (and not on the government's gold hoard) all of those crises would have been much milder, although the consequences of over-investment in railroads would still have caused losses.

There were two big Florida real estate bubbles (Cummings 2006). The first began in 1881 with a land sale by an agency of the State of Florida to a Philadelphia developer, Hamilton Disston. The State sold the land (4 million acres or 16,000 km²)—much of it the swampy everglades—at a ridiculously low price because the State owed $1 million interest on $14 million in bonds. The sale made Florida credit-worthy again, and other investors came, even though Disston only paid $100,000 in cash, the rest by note. Hamilton Disston tried to make money: He built several canals to drain the everglades into Lake Okeechobee, but the land was too flat to drain, and his project did not succeed although it left the city of St. Petersburg on the West Coast.

However, Disston's project had a positive consequence: it induced Henry Flagler, an associate of John D. Rockefeller, to take an interest in developing Florida East Coast. The project was to build a railway—the Florida East Coast Railway—and link associated real estate developments along the coast.[3] The Florida East Coast railway started from St. Augustine in 1881, reached Miami in 1904, and arrived finally at Key West in 1915. Flagler built many hotels and other facilities along the way, especially in West Palm Beach and Miami. He invested $50 million in those projects.

The land prices near Flagler's developments always rose dramatically, but the first real estate boom ended with the panic of 1893. It was also a bad hurricane year, with three big ones. The biggest, known as the Sea Island Hurricane, hit near Savannah Georgia and the storm surge killed between 1000 and 2000 people. That year was when the US Treasury almost ran out of gold, and too many people tried to convert silver-backed notes into gold. The result was that, in 1893, 500 banks around the country failed and 15,000 businesses went down with them. It ended (as mentioned already) when Grover Cleveland persuaded J.P. Morgan to borrow gold from private owners and lend it to the Treasury. That ended the panic.

There was second Florida land bubble starting after 1900 and ending in 1926. It probably started in 1904 when newly elected governor Napoleon Bonaparte Broward made a grandiose promise to "drain the everglades" (a swamp) and open up a lot of the land for development. He actually had some success. Northerners began coming, both to invest and to live. In December 1919, a man named Charles Ponzi founded the Security Exchange Company to engage in foreign exchange trade. He distributed a large number of postcards promising a 50% return on any investment in 45 days and a 100% return in 90 days. This proved amazingly seductive and such schemes are still known by his name.

---

[3] Henry Flagler (a grain salesman) and his in-laws, the Steven Harkness family, were the first financiers who loaned money to John D. Rockefeller. They became partners in Standard Oil Co. Flagler was worth $60 million when he died, after investing $50 million in his railroad and hotels.

When Ponzi was arrested on July 26, 1920—only 7 months later—he had defrauded 40,000 investors of $140 million. Not many got their money back.

Fast buck artists in Florida, called "binder boys," were inspired by Ponzi's astonishing success. One of them would pay the 10% "binder fee" on a lot coming up for sale. At the auction, he could resell the binder at a large profit. Sometimes a binder was resold several times in a day. Another scheme (also inspired by Ponzi) was the sale of waterlogged lots by mail-order. When this bubble was at its peak in 1924 and 1925, a number of northern banks, especially in Ohio, got into trouble when their depositors drew out all their savings to go to Florida (Cummings 2006).

The bubble collapsed suddenly in 1925 due to a confluence of events partly caused by its success. There was a railway embargo triggered by uncoordinated and excessive demand for shipments of goods beyond limited railway capacity. There was a ship that sank in Miami harbor and messed up real estate developer's logistics. These things had a negative impact on prices, which started to decline. Then came a bad hurricane in 1926 that left a swathe of destruction.

There were also bubbles resulting from exaggerated expectations of profits from actual investments. For instance, the boom in English canal construction reached a peak in the 1790s. In that year, only one modest project was authorized by Parliament, with a capital of £90,000, but the next 3 years it became a "mania." By 1793, there were 20 canal proposals, with an authorized capital of £2,824,700. The capital was raised by offering shares, usually in some local context, such as a church.

There are no comprehensive statistics on share prices—the London Stock Exchange did not exist until 1801—but some examples have emerged. For instance, shares for the Grand Junction Canal were issued at £100 (par). But in a single month, October 1792, they rose in price to £472. A couple of years later they were back down to par (£100). Some of those canals were never completed, and some failed, and most of that investment money was lost. In fact, there was a recession during the years 1793–97, immediately following the "mania." One factor in the recession of 1795 was the execution of King Louis XVI of France, in 1793, and the start of preparations for war, which was bad for commerce and diverted resources into armaments.

The first railway bubble was based on a construction boom following George Stevenson's successful demonstration of steam-power transport on iron rails. A number of new projects got started, and share prices rose dramatically, with spikes in 1836 and again in 1846. There was a recession after that first railway share price bubble, punctuated by a panic in 1837. That panic resulted in bankruptcies by eight of the US States (and Florida, a territory at the time). The cause was inflation (1834–37) attributable to land speculation along proposed railway lines. This was financed by borrowing from English banks. In 1836, the Bank of England tightened credit, raising interest rates and the following year the price of cotton fell 25% in a few weeks. This hurt the US planters, as well as other farmers. State revenues in the south collapsed.

Most of the states in the USA (especially Maryland and Pennsylvania) had embarked on extensive (and expensive) canal and other transportation investments, stimulated by the success of the Erie Canal and by the inflation. All of this was financed by borrowing in Europe. The panic in 1837 led to a series of defaults by US States in 1838–1841. Some of the States eventually repaid part or all of the debt, but a few did not. The Bank of England lobbied for years (unsuccessfully) for the US Federal Government to take over the State's debts. The episode was memorialized by a ditty:

Yankee Doodle borrows cash
Yankee Doodle spends it
And then he snaps his fingers at
The jolly flat who lends it
Ask him when he means to pay
He shows no hesitation
But says he'll take the shortest way
And that's repudiation.

The verse told part of the story. Of course, the investors who lost money were not entirely innocent. They were looking for a "fast buck" and got stung. For some banks in Mississippi and Florida there was no way out, and the banks failed, but most did eventually work out terms for repayment.

In the second "boom" period for railways (1844–46), new railway companies were formed with an aggregate capital of 180 million pounds, most of it invested after 1845 (Fig. 5).

Indeed, this boom consumed virtually all the financial capital available for investment at the time [ibid]. In 1845–46 at the peak of the railway bubble, 272 acts of Parliament authorized 9500 miles of new railway lines, many of them to serve the home districts of members of parliament who had invested (Colombo 2012). Share prices peaked in 1846 and began to fall, accompanied by an economic recession. From the 1846 peak to the 1850 bottom, the railway share price index fell from 160 to 60.

Most of the bubbles were about over-hyped markets. The most spectacular example was the Dot.Com bubble, 150 years later.

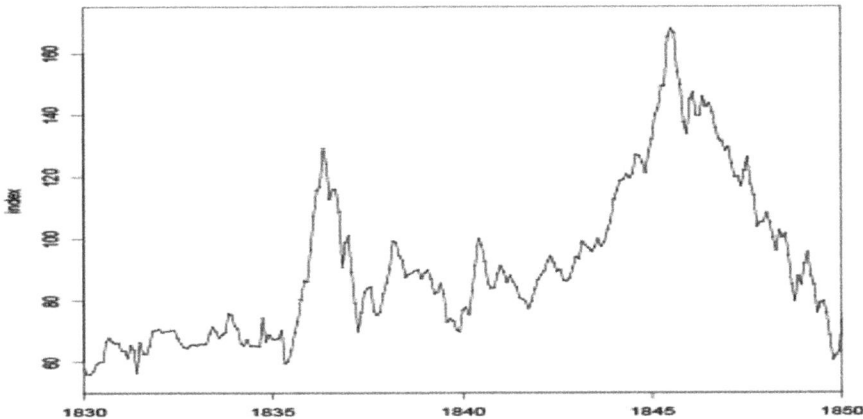

Fig. 5 Index of railway share prices in Britain 1830–1850

The Rise and Fall of Bill-Broking and the Central Bank as Lender of Last Resort (LLR)

By the end of the eighteenth century, there were many local or regional banks in the UK and most of them issued letters of credit (LOCs) from people or firms that would do business with the same bank. But the number of banks in the country increased rapidly as inter-regional trade increased (along with the canal traffic). During that period, LOCs were increasingly replaced by "bills" (or IOU's) representing present or future claims by merchants for goods or services provided by them to customers. Checks came into use around that time. Nowadays, the immediate claims by individuals are usually transfers from one bank to another via credit or debit cards. Businesses use electronic transfers for the most part. But only a few years ago, bills for small transactions were paid by paper checks on a bank account. Those checks had to be collected and redistributed by a "clearinghouse." The "back offices" of banks had to cope with payments into and out of each of the accounts of each of their depositors, and then send checks on other banks to the clearinghouse.

However, as time went on, more and more of the claims on a given depositor's account were from depositors in other banks, and if the claims exceed the current balance, payment priorities need to be established. More often than not, the deficit is temporary and the bank allows a certain over-draft to a credit-worthy client. But in difficult times—which seem to occur every few years—not all of the claims (bills) can be met immediately. There was a growing need for intermediaries who specialized in assessing the credit-worthiness of merchants and other small businesses. These intermediaries could buy and sell bills, and they could also borrow from the commercial banks, invest in safe securities (like railroad bonds?) and make small loans for specified periods on their own, providing liquidity to the money market. Such loans ("commercial paper") bridged the time gaps between sales of goods by merchants and payments to them by their customers.

In 1800, a new firm was created to specialize in the bill trade in England. The original partners, Thomas Richardson and John Overend, had been bank clerks who were familiar with the discounting trade. They saw a need for a firm, based in London, to specialize in this business. The original capital came from Gurney's Bank in Norwich. In 1806, Samuel Gurney joined the new firm of Overend, Gurney & Co. and took control in 1809. By the 1850s, the firm was a great success. Overend-Gurney (a partnership) was, during its heyday, the most trusted financial institution in England. In 1855, it had revenues of £170 million and profits of £200,000. It was much larger in terms of turnover than any of the commercial banks that existed then (Martin 2014, p. 195).

After 1815, there was a post-war recession and the BoE embarked on a "tight money" policy. The purpose was to re-establish the pre -war (1797) "specie convertibility parity" (i.e. the prices of silver and gold in pounds sterling). By 1821, specie parity was achieved and interest rates fell. Meanwhile, the breakup of the Spanish Empire in Latin America resulted in the creation of a number of new sovereign countries, eager to borrow for development purposes, but lacking in tax bases or financial infrastructure. This prompted a large number of entrepreneurial startups in England (to take advantage of the new opportunities), especially in mining or planting tobacco or cotton.

Another event in 1825 was the repeal of the Bubble Act, introduced after the South Seas Bubble in 1820. The "Bubble Act" had limited the formation of new companies to a maximum of five individual investors. This limitation must have been broken or successfully evaded from time to time (e.g. in the formation of the Stockton & Darlington Railway Company), but it greatly inhibited the financing of new enterprises. For this reason, the law was repealed in 1825. The repeal seems to have triggered (or at least enabled) the Railway Mania of the 1840s.

In 1844, the Bank Charter Act, by Sir Robert Peel's government, was issued to prevent panics such as had occurred in previous years (notably 1825, 1839) and gave the Bank of England the exclusive right to issue paper money, *but only in quantity equal to the amount of gold it held in reserve*. In principle, this restriction, the "gold standard," kept the money supply from exploding, preventing inflation, and made the English pound sterling the de facto reserve currency of the world. Japan adopted the gold standard in 1872. Other countries gradually switched from a silver standard to a gold standard in the following decades. China and Hong Kong were the last to switch, after 1908.

In 1857 the BoE did not have (and did not acknowledge) any role as lender of last resort (LLR) during a financial crisis. It did allow Overend-Gurney and the other bill-brokers to borrow from it, during the crises in 1825, 1836, 1846, and 1857, but only at the last minute and not as a matter of entitlement. In fact, the panic of 1857 was exacerbated by the restrictions in the law of 1844, and the panic was only ended when the government (of Lord Palmerston) suspended the restrictions and allowed the BoE to issue £2 million in paper currency beyond the statutory limit. This ended the panic, for the moment.

But the damage was not finished. A year later, in 1858, the directors of the BoE noticed that most of their crisis loans had gone to the bill-brokers, such as Overend-Gurney, rather than to other banks. They feared that the bill-brokers had engaged in excessively risky investments. So, in 1858, they withdrew access from the bill-brokers and closed the "lender of last resort" window. As the editor-in-chief of *The Economist* newspaper, Walter Bagehot said later, the consequences of this move were the opposite of what the directors of the BoE had intended. Instead of discouraging risky investments, the new (younger and naive) directors of Overend-Gurney made riskier investments to increase their returns. By 1860, their annual profit of £200,000 had turned into a loss of £500,000 and a capital loss of £5 million (Martin 2014). In 1861, they issued 100,000 shares of stock to the public at par £50 per share. This recouped their £5 million capital loss (Fig. 1). But the directors continued to make bad bets and the losses continued.

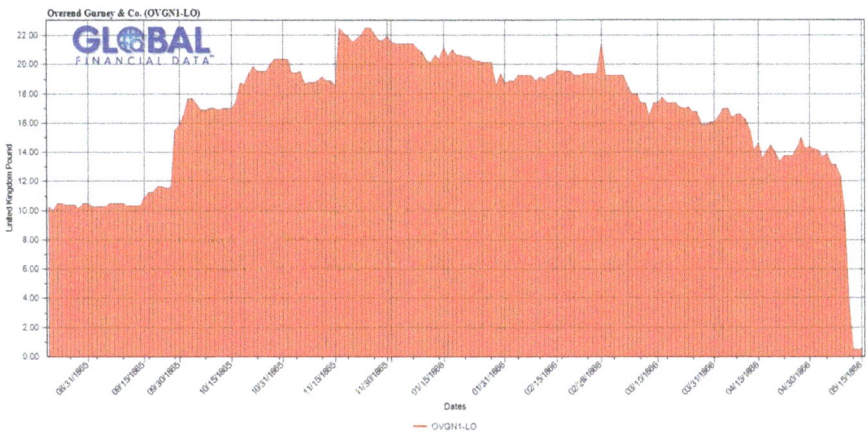

Fig. 1 The end of Overend-Gurney—share prices from August 31, 1865, to June 15, 1866

In 1865, one of Overend-Gurney's major investments, Millwall Iron Works on the Isle of Dogs, ran into trouble, resulting in £500,000 of losses for the BoE. The BoE raised interest rates in a few days from 6% to 9% and that move sank Overend-Gurney, a major borrower. On May 9, 1866, Overend-Gurney ran out of cash, and (without access to the BoE) it was forced to stop all payments to its clients. Next day—the first of the notorious "black Fridays"—there was another panic in London. It caused a run on all of the banks in the country. Only the BoE was allowing withdrawals (only by other banks), and it paid out half of its reserves in a single day.

The panic was only ended when Prime Minister Gladstone again suspended the gold-based lending limit for the BoE from the 1844 Act. That ended the panic, but not the losses. After 1866, the BoE did act as a lender of last resort (LLR), largely thanks to Walter Bagehot's critique of the 1957 and 1866 events (Bagehot 1873, Bagehot 1999). There have been quite a few financial panics and crashes since 1866, though none have occurred in Britain. The BoE successfully prevented panics by providing liquidity in 1878, 1890, and 1914 (it was nationalized in 1946).

As a matter of interest, the Panic of 1857 started in the US when the Ohio State Life & Trust Co. failed (due to a massive embezzlement). The Seamen's Savings Bank (among others) followed (see Fig. 2). That failure was incredibly contagious (to use a modern term). Within 2 months, 1,415 regional banks in the US also failed, due to runs-on-the-bank. Moreover, a number of British banks from Glasgow to Liverpool, with trade links to the USA, also ran out of gold and suspended payments of specie (gold) to depositors. Some of them

Fig. 2 Run on the Seamen's Savings Bank during the panic of 1857

also failed. In the USA, there was no central bank at the time to coordinate a response. But, in Britain, there was the BoE. The BoE played the part of an LLR at that time, even though they had no legal mandate to do so.

The Panic of 1857 has been called the first international financial crisis. It had multiple causes, apart from the embezzlement that started it. One of them was the end of the Crimean War between Britain, France, and Turkey vs. Russia. The end of the Crimean War was followed by Russia's re-entry into the global grain trade. That caused sharp falls in grain prices, which hurt US farmers and their banks. This may have been the original trigger. The discovery of gold in California in the 1840s had increased the US money supply (as prospectors spent their gold dust as money), increasing consumption and raising prices. But this increased domestic consumption also cut US grain exports. Some of the exporters and their banks also failed. In the USA, there was no central bank to coordinate a response but, in Britain, there was the BoE.

English banks started to withdraw their funds from US banks, after the failure of the Ohio Life Insurance and Trust Company. At the same time, the western land speculation bubble collapsed. This may have been partly because of the Dred Scott case (1857)—taught in all American history classes—in the US Supreme Court. That decision deprived Dred Scot—and all others of African descent—of the chance to be American citizens with Constitutional protection. It opened the possibility that slavery might soon be allowed in new states west of the Mississippi River. This judicial decision aroused public outrage and hastened the Civil War. It also adversely affected railroad bonds for projects to "open up" such western lands.

Furthermore, the gold reserve of the US Treasury had been declining as reduced US exports of grain resulted in a gold drain (English exports to the US still had to be paid for in gold). The final blow was the loss of a steam-ship, *SS Central America*, in a Hurricane off the North Carolina coast, in September 1857, with a loss of 425 lives and 14,000 kg of gold from California, intended for the US treasury. The US recession thereafter resulted in the failure of 5,000 businesses in the USA and spread to Britain, Latin America, and Europe. It lasted until the US Civil War (1861–65).

After the Civil War, the United States started a period of very rapid economic growth, accompanied by a railroad building boom, encouraged by land-grants and outright subsidies. From 1868 through 1873, 33,000 miles (over 50,000 km) of railroad tracks were laid, mostly west of the Mississippi. The financing of railways at the time was largely organized by one investment bank, Jay Cooke & Co. (which had previously been the seller of War Bonds to finance the Civil War). However, in 1873 Jay Cooke over-reached by financing the building of the Northern Pacific Railroad in competition with the Union Pacific railroad. I will come back to that later.

There was a financial panic in the USA in 1873, followed by labor problems, and a major recession that lasted until 1879. This one may have started in Europe with the German demonetization of silver in the aftermath of the Franco-Prussian war of 1871 and consequent bank failures in Vienna. The lower international price of silver caused some distress among the US silver miners. President Ulysses S. Grant signed the Coinage Act of 1873 (also known as the Mint Act of 1873 or the Fourth Coinage Act), the ramifications ended up being so notorious that many would later call it the "Crime of 1873." Briefly, it recommended moving the United States off of bimetallism in favor of the gold standard. In 1867, diplomats in Paris discussed turning away from silver toward a gold standard to help standardize currency across international borders. This "be kind to miners" Act, that demonetized silver actually allowed the silver price to rise. That, in turn, increased the demand for gold, which was now the only basis for money. (The Chicago fire of 1871 and the great Boston fire of 1872 were other contributory factors to the Recession.)

But, in retrospect, the primary cause of the recession of 1873–1879 was not demonetization of silver, but overbuilding of western railroads. In October 1873, Jay Cooke & Co. failed, with massive repercussions. A month later, 55 railroads had failed, and a year later, 60 more railroads had failed, along with 18,000 businesses. Unemployment reached 25% by 1875. There was a second recession in 1877, due to the first great railway strike. In his landmark book, "Progress and Poverty," Henry George attributed the recession of 1873 primarily to real estate speculation, which he believed to be the primary cause of "boom-bust" (George 1879). His advocacy of the land tax (as a substitute for other taxes) was strongly influenced by that experience.

There was another US financial panic in 1884 (in the middle of a recession) with three important bank failures, and 10,000 business closures. The panic of 1884 was triggered by a European gold shortage and actions by New York Banks halting investments in the United States. Another panic, "the Baring crisis," followed in 1890; it was triggered by bad (British) investments in Argentina.

The Sherman ('be even kinder to silver miners') Silver Purchase Act of 1890 committed the US Treasury to buy 4.5 million ounces of silver per month from the western silver mines *at market prices*. Payment had to be in gold. This, in turn, almost totally depleted the US treasury (of gold) and also caused the price of silver to fall. By the time Grover Cleveland took office in 1893, the US Treasury gold stock had fallen below $100 million for the first time since 1873. And the decline accelerated.

The Panic of 1893, like the panic of 1877, was also caused by railroad overbuilding. It started with the failure of the Reading Railroad. This, in turn, led to the failure of thousands of small businesses and hundreds of banks in the area served by that railroad. The local recession morphed into a major

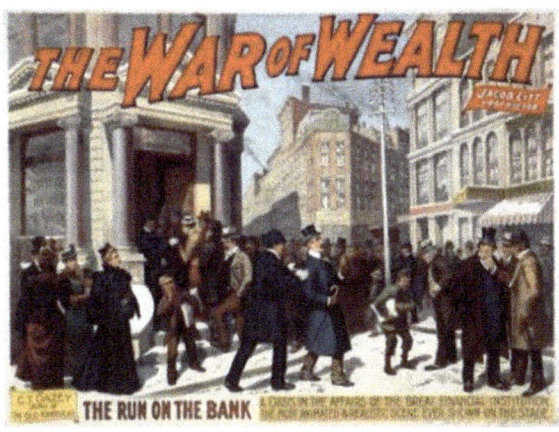

Fig. 3 Left, frenzied stockbrokers on May 5, 1893. Right, a 1996 melodrama inspired by the 1893 crisis

nation-wide depression that spread to Europe and did not end until 1897 (Fig. 3). US unemployment reached 4 million at its peak.

Unfortunately the supply of gold in the US Treasury was so depleted by the Sherman Silver Purchase act that US economic activity nearly ceased for lack of money. The Sherman silver purchase act was repealed in that year. Many silver mines were closed in consequence, along with 15,000 businesses and 640 banks.

But the panic of 1893 also caused a lot of people to hoard gold. It was effectively a "run" on the US Treasury. By 1895, the federal government was literally almost out of gold. That would have prevented the US government from paying its bills. President Cleveland asked J. P. Morgan to organize a consortium to lend $62 million in gold to the treasury. (It is alleged that this gold was provided by the Rothschilds.) This maneuver enabled the treasury to issue $7 million (*$200 million in today's money*) in "gold certificates" to its creditors. That gesture enabled the federal government to avoid default. Sometimes a gesture is all that it takes.

President Grover Cleveland was heavily criticized for this move by populist William Jennings Bryan, a fellow Democrat, who advocated using silver as a basis for money—because there was more of it—and complained in a famous speech that farmers were being "*crucified on a cross of gold.*" This phrase meant that the gold shortage (thanks to demonetizing silver) had made it too difficult for farmers to borrow from banks for current expenses. Bryan wanted to remonetize silver. Consequently, Cleveland lost the 1896 election to Republican William McKinley, who was strongly supported by J. P. Morgan. McKinley (like Morgan) favored the gold standard.

Some have said that William McKinley was a mere puppet, entirely controlled by the great financial magnates. (The *Wizard of Oz* was a political satire; the "yellow brick road" was a fairly obvious reference to gold politics (Littlefield 1964).) But others say that Morgan himself was merely a Rothschild agent (Kolko 1963; Mullins 1983). The need for the US Treasury to borrow gold in 1895 from private bankers was ironic since the United States was actually accumulating gold at the time, but it was all in private hands.

The nearest US counterpart to the BoE in New York, before the Federal Reserve Bank was created (in 1912), was J.P. Morgan and partners. But while the US Federal Reserve Bank (a consortium of 12 privately owned regional banks) was designed to be a lender of last resort (LLR), it has rarely performed that function (Humphrey 1989). Instead, the Fed has focused almost exclusively on monetary policy.

The financial history of nineteenth century Britain is shown in Fig. 4, compiled by Simon Kuznets. The quasi-periodicity is obvious. The 4-year Kitchin inventory cycle is superimposed on a longer and more irregular investment cycle with much bigger swings. The deep bottom in 1857—already described—was triggered by a tidal wave of bank failures in the USA which started in Ohio and spread to England. It had multiple causes, including declining grain prices (due to the Russian re-entry into global markets after the end of the Crimean War). Another cause was a gold shortage in the USA that adversely affected farmers. That recession in the USA had world-wide repercussions.

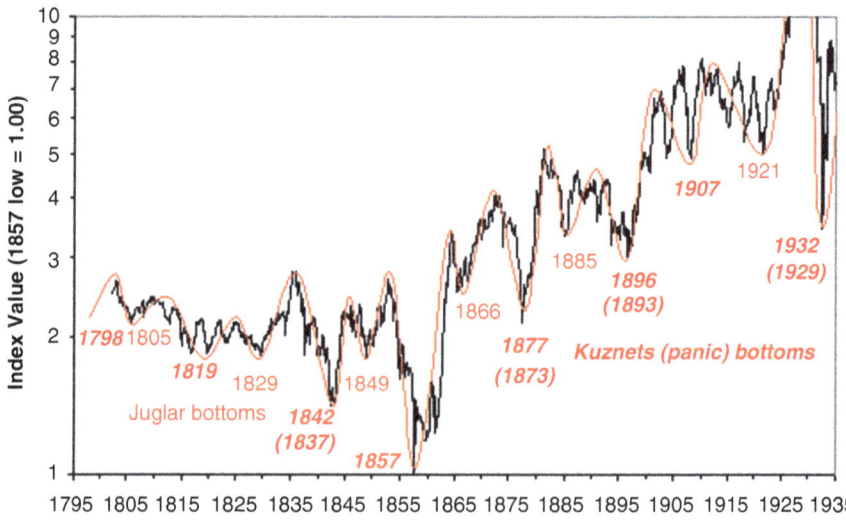

Fig. 4 Financial history of nineteenth century Britain: 1795–1935

The Rise and Fall of Bill-Broking and the Central Bank as Lender... 133

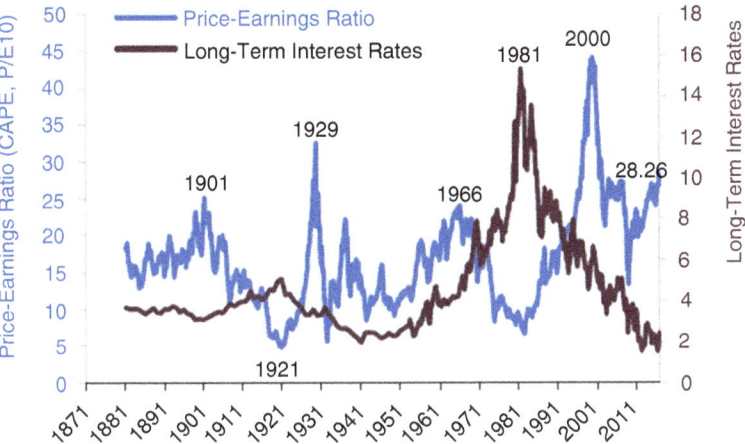

Fig. 5 The Shiller housing price index vs. long-term interest rates

A similar pattern of periodicity with short cycles imposed upon longer ones can be seen for the financial history of the USA since 1860. Figure 5 was compiled by Nobel Laureate Robert Shiller, of Yale, the author of the book "Irrational Exuberance" (Shiller 2006). In this analysis, it is clear that one of the drivers of investment is the long-term interest rate. This makes sense because the lower the interest rate, the easier it is to borrow money via mortgages. Interest rate peaks coincide with the prices of stock market shares. Again, the cyclic pattern is obvious, albeit the periodicity is less regular.

Why the periodicity of major events? The simple answer may be that after an avalanche, the snow-pack is depleted. The crisis has ended the imbalance that caused it, so there will be a delay before the next imbalance becomes critical. This makes sense, whether the imbalance is due to excess debt or over-investment in some project. Figure 1 in chapter "A Brief History of Financial Booms and Bubbles" show the correlation between housing prices and long-term interest rates. When interest rates are low, prices are high, and vice versa.

Not all the important financial crashes were in the USA or Europe. The Xinhai revolution in China occurred in 1911 (see the map (Fig. 6) and the photo (Fig. 7)). The Qing dynasty of Imperial China was replaced in that year by a republic, under the leadership of Sun-Yat Sen, and his Kuomintang (Republican) party. The details would require many pages to summarize, but one consequence was the spectacular failure of the largest banking group in the country, the Piaohao group, centered in Pingyang, Shanxi province.

That network of banks was founded in 1836, when the Xiyusechang Dye Company converted itself into a bank and expanded rapidly. By 1900, there were 32 interlinked banks with 475 branches covering most of China and some of Indo China. The Piaohao banks were important cogs in the wheel of

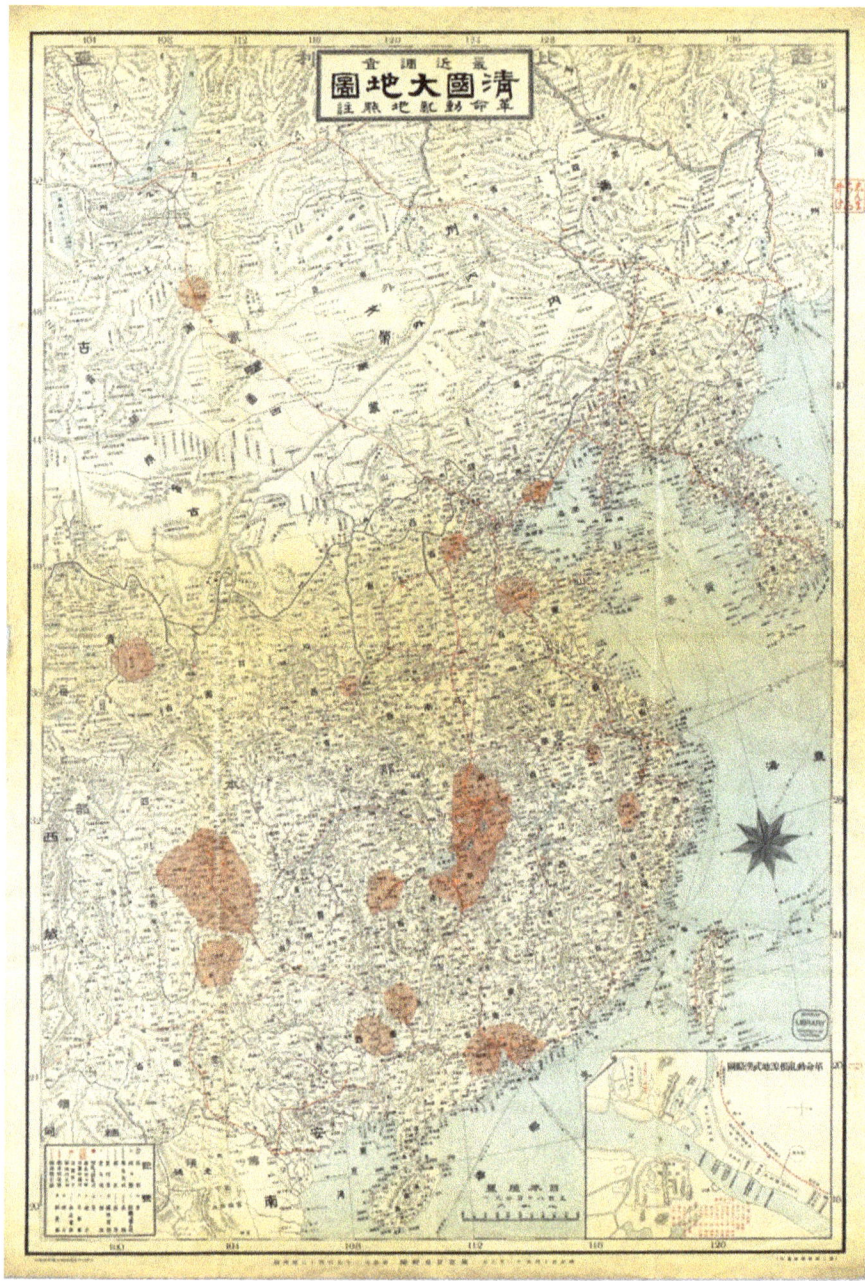

Fig. 6 Map of the uprisings during the Xinhai Revolution

The Rise and Fall of Bill-Broking and the Central Bank as Lender... 135

Fig. 7 Nanjing Road, Shanghai, after the revolution, displaying "5-races under one union" flags, used by the revolutionaries

commerce. There were, at the same time, a large number of small local banks called qianzhuang, mostly individuals performing exchange services, similar to the bill-brokers (like Overend-Gurney in England) but not part of an integrated network.

All the Piaohao banks failed simultaneously in 1911 because they had loaned significant sums of money to the Empress (who was regent for the 6-year-old Emperor), and because neither the Xinhai revolutionaries nor the Imperial Government were willing to pay her debts, which they chose to regard as personal. The lack of a banking system almost certainly contributed to the failure of the Kuomintang government—largely dominated by military men and civil servants—to achieve any significant economic recovery in the following years. That failure, in turn, opened the door to Japanese aggression in the 1930s and the victory of the Communists under Mao Tse-Tung after WW II.

The year 1901 saw the first major US stock market crash. It was triggered by the fight between E.J. Harriman of Union Pacific (and Southern Pacific), Jacob Schiff (Kuhn Loeb & Co.) and James J. Hill (allied with JP Morgan) for control of the Northern Pacific Railroad. Hill wanted to purchase the Burlington-Northern line (to connect the Twin Cities with Chicago) and Harriman—who had interests in a competing railroad—wanted to prevent it.

The competitive purchases of Northern Pacific shares by Harriman and Schiff on one side and Hill and Morgan on the other essentially cornered all the shares between them, drove the share price sky high, and incidentally ruined a number of "shorts" (speculators who have borrowed shares of stock and sold it, hoping to repay later when the price goes down). Every bubble leaves casualties on both up and down sides. Of course, the "bubble" eventually collapsed and that hurt many other investors. The market gyrations were felt in both London (where much of the money on both sides originated) and New York.

The panic of 1907, that followed, had a significant effect on the future of the US electrical industry. It was a financial maneuver that got out of control and caused enormous harm, including many bank failures and corporate failures. The trigger was an attempted "short squeeze" by one Otto Heinze. He was the brother of F. Augustus Heinze, who controlled the "United Copper Company" of Butte Montana. United Copper was fighting Anaconda Copper, on the basis of its ownership of a small piece of land on the top of the hill where Anaconda was mining. Their lawsuits were based on the "apex" theory of ownership of underground resources. United Copper had won twice in local courts in Montana and had forced Anaconda to stop operations. This caused massive unemployment locally, but its stock was "in play" by speculators.

Otto Heinze's idea was to raise the price of the stock of United Copper (then $39 per share) by buying more of it aggressively (with borrowed money). The price rose to $60 at peak. Otto Heinze hoped that he could force the "shorts" to buy his newly purchased shares, at much higher prices, to replace the ones they had borrowed. As it turned out, he misjudged the market badly. The "squeeze" did not work. The value of the United Copper shares he had used as collateral for the loans then dropped sharply, from $60 down to $10 almost overnight. He (and his brother) were both ruined.

But it did not stop there. His brokerage firm was also ruined because he could not repay the money he had borrowed. The Heinze bank in Butte Montana was ruined. United Copper, itself, never recovered and Anaconda took over its assets. Firms using United Copper stock as collateral were ruined. They in turn had borrowed from other banks, which were also ruined.

Because of this domino effect, the Knickerbocker Bank and Trust Co., the third largest bank in New York, also failed. The panic of 1907 dried up liquidity and created a "squeeze" on many corporate borrowers. One of them was George Westinghouse, principal shareholder of the Westinghouse Electric manufacturing Company. Westinghouse Electric, while basically sound and growing fast, had built a 22-mile tram-line (to demonstrate the technology) that was currently unprofitable. He had included shares of infant startup subsidiaries as part of its corporate financial reserves. It had also borrowed a lot of money for short-term operating expenses, using its stock as collateral for the loans.

A year earlier, Westinghouse Electric could have sold new shares easily, but it did not do so (Carpenter Jr. 1916). General Electric Company, Westinghouse's competitor, had done the opposite, selling new shares and cutting debt. The Westinghouse Company was suddenly bankrupt and in receivership by October 1907. George Westinghouse—one of the greatest entrepreneurs of all time—had used his own shares as collateral for the loans, so he was financially ruined. The lenders took over. Westinghouse Electric Company recovered quickly under new management. But that is why there is no "Westinghouse" fortune, in case you were wondering. (You may also wonder why there is no Edison fortune. But that is another story.)

Bubbles and Panics Since 1920

To finance their rapid growth in the postwar period, many companies issued shares, which were traded on the New York Stock Exchange. During the 1920s, demand for short-term corporate loans declined substantially, possibly due to the accumulation of profits during the war, together with unprecedented growth of the retail stock market. During this period, the big banks increased their holdings of equities and long-term bonds. Those holdings constituted a time-bomb that contributed to the events of 1929–1930.

A financial innovation of the pre-1929 period was the highly leveraged *investment trust*. Investment trusts were initially created back in the 1880's specifically to invest in operating companies, on behalf of a group of investors. The trust was mainly a device to keep control over a number of companies in the hands of a few financiers. Such trusts only invested in stocks, not in real productive assets. (The first example was the Standard Oil Trust, mentioned earlier.) Trusts were able to sell their own shares to the public. This left all decision-making authority in the hands of their sponsoring investment banks. The sponsoring banks were paid a management fee as well as receiving income from the purchase and sale of trust shares.

The business model of the investment banks in the 1920s was to issue shares of existing companies or leveraged "investment trusts," while the commercial banks promoted margin loans[1] to brokers. The commercial banks

---

[1] A client could purchase 100 shares of XYZ Corp for the price of (say) 25 shares, the remainder being financed by the brokerage (using money from the banks) and held as security. If the value of the shares went up, the customer kept the profits on all the shares. However, if prices went down, the broker would make a "margin call" asking for more security. If the client could not find more money, the broker would sell shares.

made their profits from lending to brokers based on the "rediscount rate" set by the Federal Reserve. Speculative margin buying increased the demand for shares, which drove up share prices. Rising corporate profits (which were quite real) also led to rising share prices. This encouraged still more margin buying, much of it in so-called bucket shops operated by unscrupulous fast-buck artists, to take advantage of the ignorant "suckers," of which there were many.

The number of such investment trusts in the USA before 1921 has been estimated as about 40, of which US Steel (the mega-merger organized by JP Morgan) was the most outstanding. But the number grew much faster than the prices of stocks. There were 160 such trusts at the beginning of 1927 and 300 or so a year later. In the year 1928 alone, 186 additional new trusts emerged, and in 1929 another 265 were created. The value of shares held in such trusts during 1927 was $400 million. In 1929, the NYSE finally allowed the trusts to be listed. The value of shares in investment trusts sold to the public reached $3 billion in that year, which was at least a third of all capital funds raised during the year. By the time of the crash in October, the trusts had total assets of more than $8 billion (Galbraith 1954, pp. 45–55).

Some investment trusts employed leverage, usually by selling bonds and preferred stock as well as common stock. In these cases, a rise in the value of a common stock in trust A would increase the value of A since the value of the bonds and preferred shares would not have changed. It would also increase the value of trusts B and C that held shares of A. Thus the gain is multiplied. This fact enabled speculators to multiply their paper gains in trusts by investing in each other's trusts.

A series of activities by Goldman Sachs, beginning on December 4, 1928, epitomizes what was going on (Galbraith 1954, pp. 66–67). Actually, the driving force was a senior partner named Wadill Catchings. On that day, the firm issued shares for an investment trust called Goldman Sachs Trading Corporation, with an initial capital of $100 million. Goldman then resold 90 percent of the shares to the public for a total of $93.6 million leaving it with a net investment cost of $6.4 million and total control of a subsidiary with $100 million in cash at its disposal. Goldman Sachs Trading Co. then engaged in a share buyback of 560,724 of its own shares in the market (over half) at a cost of $57 million, leaving $43 million in cash.

Meanwhile the market price of the remaining shares outstanding rose from the original $104 to $222.50 by February 7, 1929. On February 21, Goldman Sachs Trading Company acquired another investment trust called Financial and Industrial Securities Corporation, bringing the combined nominal assets of the Goldman Sachs Trading Company to $235 million. Some of those shares it sold to William Durant (the founder of General Motors) who then resold them at a profit, driving prices still higher.

Then in the summer of 1929, the Goldman Sachs Trading Company, in partnership with investor Harrison Williams,[2] launched two new investment trusts, by name Shenandoah and Blue Ridge, within a month. The total stock sales for these two trusts were, respectively $177 million for Shenandoah and $142 million for Blue Ridge, for a total of $319 million (ibid.). The trading company still had most of its $235 million in cash in the bank, minus a few million for expenses.

This money was then used to buy small companies at inflated prices. But prices stopped rising and started down in October 1929. By 1931, losses by Goldman Sachs Trading Company amounted to 70% of all the losses by 14 leading investment trusts (Ellis 2009, p. 28). A few years later, the value of the Goldman Sachs Trading Company had fallen from a peak above $250 per share in February 1929 to a low of $1.75 per share.[3] Leverage works both ways. It is worth noticing that Goldman Sachs, the investment bank (a partnership) that created and sold the Trading Company, was not significantly affected. Only its clients, the stockholders, lost their money.

Leverage through trusts magnified the monetary losses when buying enthusiasm waned and prices began to fall. By October 1929, there was over $8 billion in broker's loans outstanding. The loans to the brokers constituted a significant fraction of the assets of many banks. These "margin calls" drove prices down and triggered still more margin calls. The "crash" in October 1929 led to a flood of margin calls as brokers sold into a declining market. That triggered what has been called "a race to the bottom." It hurt many brokers and banks, as well as millions of ordinary citizens.

The 1929 stock market crash was essentially ephemeral, in that it was driven by an imaginary future in which everybody can get rich simply by investing in stocks. The crash, potentiated by wrong-headed government policies, triggered the Great Depression. In fact, the immediate and longer-term losses far exceeded any benefits from the boom. Even the "winners" (such as the Hollywood movie industry) did not create much real wealth in comparison to what was lost.

In 1980, there was a brief financial panic, quite similar to the one in 1907, but luckily contained (Streeter 1984). It was caused by a nearly successful attempt by the Hunt brothers (sons of H.L. Hunt, of Texas oil fame) to "corner" the silver market. They bought a large amount of silver—estimated to be

---

[2] Harrison Williams was reputed to be the richest man in the USA in 1929 having made his fortune by organizing Central States Electric Co. He had a partnership with Wadill Catchings.

[3] For a detailed description of the disaster, which nearly destroyed Goldman Sachs, see (Ellis 2009, Chap. 2).

a third of all the silver not owned by governments—on margin. In this endeavor, they used the services of the brokerage firm Bache, Halsey, Stuart & Shields (later Prudential-Bache Securities, now simply Prudential Securities) in which they held a 6.5% share that they failed to tell the Security and Exchange Commission (SEC) about. In January 1979, the market price of silver was $6.08 per Troy ounce. On January 18, 1980, the market price of silver was $49.45 per ounce, an increase of 713%.

The rising prices caused serious problems for jewelers (Tiffany's published a full-page ad complaining about the Hunts) and very serious problems for the photography industry which was the main consumer of silver. Kodak, Agfa-Gevaert, Dupont, Ilford, and Chemco were all hurt in various degrees. After the peak in March 1980, the price dropped (as always happens after a bubble peaks) and fell 50% in 4 days. This triggered a margin call of $100 million, which the Hunts could not meet. At that point, the silver market collapsed and the broker faced ruin.

Unlike the 1907 case, the big New York banks put together a $1.1 billion loan package to the Hunts (collateralized by other properties), enabling them to pay what they owed, thus rescuing the brokers and the banks. The Hunt brothers were still worth $5 billion at that point, but by the end of the decade, their wealth was down to $1 billon, thanks in part to adverse lawsuits from companies that had been hurt.

Incidentally there was a peak in the price of gold at the same time as the peak in the price of silver, reflecting the close historical relationship between silver and gold prices. The speculators buying gold at that time presumably assumed that the rising price of silver was caused by the rising price of gold. They got it backwards, but it did not matter. Gold prices fell when silver prices fell (Fig. 1).

In 1987, there was another financial panic on Wall Street. The Dow Jones Industrial Average (DJIA) had previously risen 44% from the end of 1986 to an all-time peak in August 1987, but had started to lose ground in September. That "bounce" was generally attributed to the market response to the collapse of OPEC in early 1986 and the sharp (50%) drop in oil prices that followed. On Thursday October 15, a "Silkworm" missile from Iran sank an American owned (Liberian registered) Supertanker off of Kuwait's main oil port. The next day, another American owned (and flagged) tanker was hit by another Iranian Silkworm. By the end of the day, the market had already lost 12% of its August peak. It occurred on "Black Monday," October 19, having started hours earlier in Hong Kong and Australia. The DJIA fell 508 points (22.6%) in 1 day, while the Standard and Poor (S&P) 500 lost 60 points (21%) (see Fig. 2). Quite a lot of people blamed computerized program trading (new at

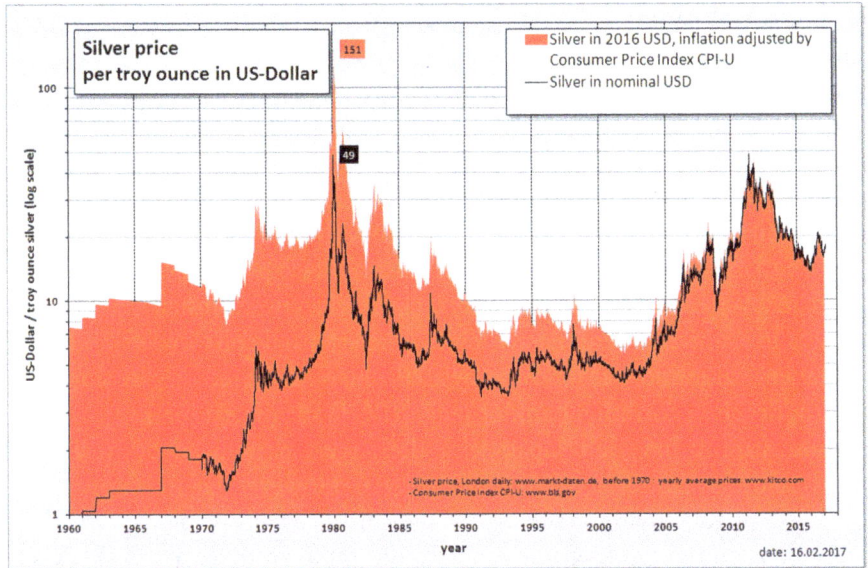

Fig. 1 The great silver bubble of 1980

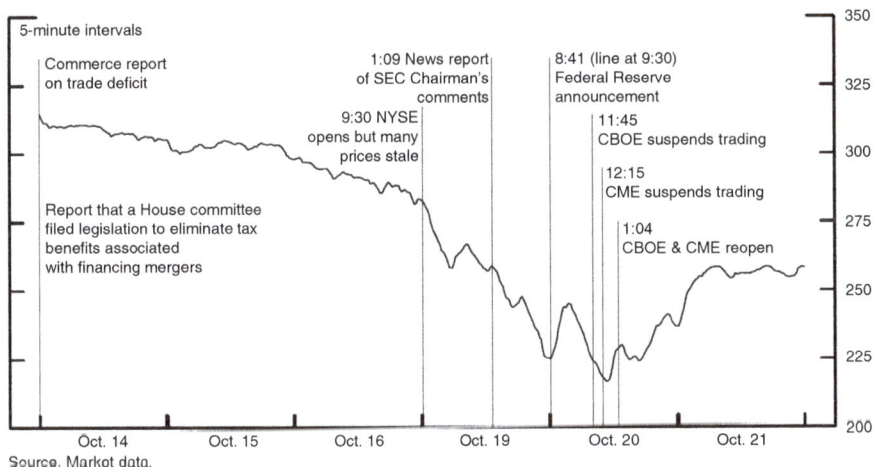

Fig. 2 The behavior of the S&P 500 around the time of "Black Monday"

the time) for the severity of the crash. Other factors were certainly involved, and it is unlikely that any single one of them was responsible.

The Black Monday incident was unusual in several ways. It was not associated with any bank or financial difficulty. There is speculation that a small number of computerized program traders may have caused the problem. Other factors have also been suggested. It was the first time the Fed intervened directly as Lender of Last Resort (LLR). Fed Chairman Greenspan urged the banks to continue lending to brokers (possibly recalling the silver panic 7 years earlier) and they did. It was over quickly. During the next two sessions of the market (October 20 and 22), the market recovered 57% of the one-day loss. In fact, the S&P set new records less than 2 years later.

The Japanese land-stock bubble of 1986–1992 (see Fig. 3) was caused by land scarcity for building in Tokyo. The photo (below right) is the Tokyo skyline. The land price bubble morphed into a stock-price bubble and also a boom in overseas investments, where land prices were much lower. Those overseas investments were mostly losers. The eventual collapse left nothing of permanent value and was followed by a 20-plus-year period of very slow GDP growth, despite enormous Bank of Japan (BOJ) stimulus efforts.

Those stimuli have left Japan with the largest national debt (to its own banks) in the world, 2.504% of GDP in 2016. However, Japan is now canceling this debt (by buying it back from the banks (using "printed money" created by the central bank) at the rate of $720 billion per year) and there is no evidence of any inflationary consequence in Japan. The fact that this massive bank debt is being reduced at no visible cost to individuals or to corporations is a fact worthy of serious consideration on the part of economic theorists.

The Dot.Com bubble of 1997–2001 was partly triggered by interest in fast-growing internet companies, and partly initiated by changes in the US tax laws, viz. the Taxpayer Relief Act of 1997. That change took capital gains rates from 28% for long-term investments, down to 20%. This favored capital gains over dividends, for many high-income investors. That shift in preferences, in turn, probably motivated an increase in investment in equities vs. bonds, ceteris paribus. The US "dot.com" bubble of 1997–2003 was big and was not contained by the Fed. It was surprisingly similar to the railway bubble in the 1840's in England.

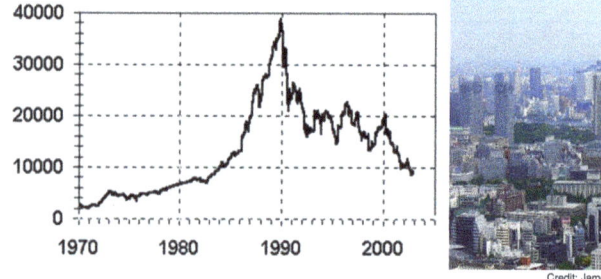

Fig. 3 Nikkei stock index and Tokyo real estate

The main factor that drove the "dot.com" bubble was a brief hysteria over the potential of the future of the "internet" itself. That was a newly hatched term for the international network of computerized information systems (or the "information superhighway" as Al Gore famously called it). Business publications enthused over the cable revolution, and "interactive TV" with 500 channels, and "do it yourself" entertainment (games) and news. *Wired* magazine was started in 1993.

Meanwhile the World Wide Web, invented by Tim Berners-Lee in 1989 (at the international high energy physics (CERN) laboratory in Switzerland), was growing like an algal bloom. Protocols for creating and operating "websites" were being developed. The number of websites increased explosively during the early 1990s. There were 1 million "edu" websites and another million "dot.com" commercial websites in the year 1994. That early growth was fueled by easy access to the copper-wire-based telephone lines everywhere, which were initially quite underutilized. Glass fiber cables were great for long distance communication, but they were not economic (yet) for "the last mile." There was a major debt-financed, over-investment in optical fiber cable capacity in 1989–90 (part of the reason for the later demise of WorldCom and Enron). It was all about digital technology replacing analog technology.

By 1990, the Internet was already growing fast (300,000 nodes in that year) and non-academic users were beginning to get access to it. America On-Line (AOL) began in 1989 as a specialized dial-up service for Apple Mackintosh users. Soon AOL began distributing free floppy-disks, and later CD-ROMs. Marc Andreeson at the University of Illinois created a successful browser (called "Mosaic") that was acquired and commercialized by AOL as "Netscape." AOL had a million subscribers by 1994.

There were lots of startups for web-based services. They were doing well until 1995. In that year, Bill Gates at MicroSoft demanded a share of Netscape's browser business on favorable terms "or else" he would put Netscape out of business. Jim Clark (the founder) of Netscape got mad and said "no way," and Gates proceeded to do what he had threatened. He did it by creating a rival browser called "Explorer" and "bundling" it into MicroSoft's "Windows '95." Bye bye AOL. Clark then complained to the Federal Communications Commission (FCC), but got nowhere.

The US Telecommunications Act of 1996 was supposed to force the local telephone companies to open their copper wire-based networks to all users. The existing second generation (2Gto) system was supposed to painlessly become a third generation (3G) system, with some glass fiber connections.

The glass fiber industry boomed in the 1990s, as a number of companies saw a great future for data transmission on the "information superhighway."

A lot of new companies called Competitive Local Exchange Companies (CLECs) were formed to provide this transition service. But the old-line telephone companies or Independent Local Exchange Companies (ILECs) refused to allow open access, despite the law. The CLECs sued the ILECs in court. But the case was not decided until 2002, which was too late for the CLECs. Most of the investment that went into them went subsequently down the drain.

The "Dot.Com" bubble got its name from the fact that a number of "high flyers" from Silicon Valley were fast-growing startups with rosy business prospects, based on the analog-to-digital transformation throughout information technology (IT), but having no current profits. Some were the exemplars of a new kind of "platform" company with positive returns to scale, confounding conventional economic wisdom (Arthur 1988). Skeptics were dismissed by enthusiasts at the time as "not getting it." But when share prices stopped rising, most of the startups collapsed.

Among the dot.com stars were two major frauds, Enron and WorldCom. Both were elaborate constructs, based on a series of mergers and acquisitions using borrowed money, with the founding company's own stock—ever rising in price—as collateral. WorldCom started as a merger organized by Bernard Ebbers between two small local telecom firms in Mississippi. It acquired more and more assets (with borrowed money) and finally acquired Microwave Communications Inc. (MCI) in 1998. Then the combined firm (WorldCom) proposed a merger with Sprint (originally General Telephone & Electronics Corporation [GTE]), which would have been the largest in US history. But that merger was resisted by the US Dept. of Justice on anti-trust grounds. That delay caused the market value of the stock to decline, which triggered WorldCom's financial difficulties and bankruptcy. (MCI now survives as a unit of Verizon.)

The Enron case was similar, in that financial fraud was also involved. Enron started as a merger, driven by Jeffrey Skilling, between two small local natural gas companies in Houston in 1985. Fifteen years later it was a giant, with 29,000 employees and interests in gas transmission, electricity, pulp and paper, and telecom, with revenues around $100 billion. For six straight years, it was the rated "America's most innovative company" by Fortune Magazine. But it turned out that the company had avoided taxes by means of complex "off balance sheet" accounting devices. The discovery of that fraud destroyed its auditing and accounting firm, Arthur Anderson & Co. It triggered a major regulatory change, the Sarbanes-Oxley Act of 2002. Both Ebbers of WorldCom and Skilling of Enron went to jail.

The frauds at ENRON and WorldCom made matters worse for the other Dot.Coms. The bubble created and destroyed a great deal of wealth. The market lost 70% of its value during the next 18 months after March 2000. Those losses were not "acts of God." They were caused by actions of greedy capitalists. A lot of the losses were suffered by people who had nothing to do with the frauds (Fig. 4).

The four biggest gainers in the Dot.com stock market, and their peak price-earnings (PE) ratios, were Microsoft (59), Cisco Systems (179), Intel (126), and Oracle (77). In recent years, all of those companies are still around, but their PE ratios have dropped (as of 2017) to 20, 13, 17, and 16, respectively, which is slightly less than the average for the S&P 500. The bubble did leave some internet companies that "went public" during that period (including AOL, Yahoo, eBay, and Amazon), but companies that waited (like Google) were able to raise money during the subsequent downturn.

This brings us to the financial crisis of 2008. From one perspective, that of monetary economists, the cause of the crisis was the Fed's over-reaction to the Dot-com crisis of 1999–2000, namely ultra-low interest rates for too long. What that low interest encouraged was excessive investment in real estate, especially in southern California, Texas, and Florida.

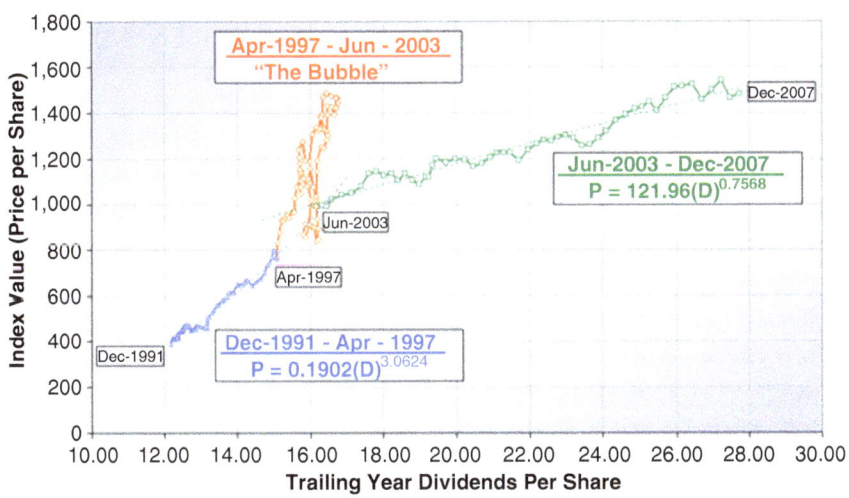

Fig. 4 S&P stock index plotted against per share dividends

From a more technical perspective, the cause of the 2008 crisis was a shift in bank behavior that started in the 1990's. There were several factors. One factor was that the commercial banks had swallowed up all the independent investment banks or, in a few cases, the investment banks (e.g. Goldman Sachs) became a commercial bank, after the demise of the Glass–Steagall legislation from 1933. The old partnership structure of investment banks had been replaced by a limited liability corporate structure, in which the directors are not personally "on the hook" for the company's investments, whereas a partner is personally liable. The commercial banks no longer have "skin in the game." They make money by offering services for a fee, and creating and selling investments to other investors, like insurance companies and pension funds. Home mortgages were less and less part of a bank's reserves. They are increasingly "hived off" to other investors.

Another factor was that, back around 1970, some people at Salomon Brothers figured out how to package a "bundle" of home mortgages into a bond that could be sold to a long-term investor, with a fixed interest rate and a fixed redemption date. Those bonds became very popular among long-term investors, like pension funds, so the Wall Street banks made creating and selling them a big part of their business. Once upon a time, a would-be homeowner would go to a local bank or Savings and Loan Association to negotiate a mortgage loan. Now most mortgages have to be insured by the Federal Housing Authority (FHA) and pre-financed by the Federal National Mortgage Association (FNMA) known as "Fannie Mae."[4] Mortgages are created by private companies that work with real estate agencies. They advertise for customers, do the paperwork, get the financing (usually from FNMA), then sell the mortgages to a bank.

The bank then packages them into mortgage-based bonds (called Collateralized Debt Obligations, or CDOs), which are quickly resold to pension funds or insurance companies. The "packaging" process takes a few weeks, so the mortgages were in the bank's possession only for that time. Nobody knows how many CDOs were sold in the run-up to 2008, but, in nominal terms, it was in the trillions (not billions) of dollars.

The problem was that the mortgage companies (and the banks) did not care if the buyers could not pay because they were not planning to hold onto the bonds. They were selling houses to anybody who would walk through the door, including NJNAs (no job, no assets). They offered adjustable rate

---

[4] FNMA is a private company traded on the stock exchange but with official government backing. This reduces its riskiness according to the Bank of International Settlements (BIS) in Basel, which makes such determinations.

mortgages (ARMs) with zero interest for the first 2 years, and sometimes even with cash back. Needless to say these "bonuses" were used to induce unqualified buyers to sign up. The costs were added on to the total price of the house. Like all bubbles, it worked beautifully as long as prices were rising. It fell apart quickly when prices peaked and started to fall. Then the mortgage defaults started to increase radically.

When the default rate on the bonds rose above the "normal" level (based on earlier data from the time when people with no money were not buying houses), the AAA ratings were down-graded. They soon fell to "junk" status—below "investment grade". At that point, the CDOs could no longer be used as part of the capital base for financial institutions, nor could they be sold to pension funds. In fact, the CDOs became un-marketable because nobody knew what they were worth. All the banks had counted unsold CDOs as part of their capital reserves, but in mid-2008, they could no longer do so. Funds that contained CDOs were suddenly worthless, or at least non-liquid. Bear Stearns nearly failed for that reason. (It was rescued by a consortium of big banks led by the New York Fed.) This meant the banks "leverage" rose, which adversely affected their market value. Most of the mortgage companies, such as Country-Wide, were suddenly out of business.

The banks thought they had "insured themselves" by buying derivatives called Credit Default Swaps or CDSs for their CDOs. These CDSs were supposed to pay off if, or when, the insured CDOs lost their value. The CDSs were sold primarily by American International Group or AIG, the country's largest insurance company. AIG had underpriced them, thinking they were money for jam, and had no reserves in case of need to pay off. Lacking a lender of last resort (LLR), the US government created a Troubled Asset Relief Program (TARP) with $800 billion to spend in the fall of 2008.

Many Congressmen must have thought the money would be used to help homeowners avoid foreclosure. No such luck: the money all went to bail out the banks and a few other firms that were designated as "too big to fail" (TBTF) and that had to be rescued, because without bailouts, the whole economy might fail. The two biggest bailouts went to AIG itself ($185 billion) and GM Capital ($130 billion). It is not surprising that when the bailouts came through, Goldman Sachs, one of the biggest beneficiaries of the sub-prime game, was first in line for payment (from AIG) on its CDSs.

The financial losses on Wall Street affected the stock market adversely, as shown in Fig. 5. The graph for the S&P is virtually identical, adjusting for the scale. Stock market losses in 2008–09 were very large; almost 50% of the market value of shares in 2008 was destroyed in a few weeks. (Most of that loss, and more, has subsequently been recovered, but the recovery has been

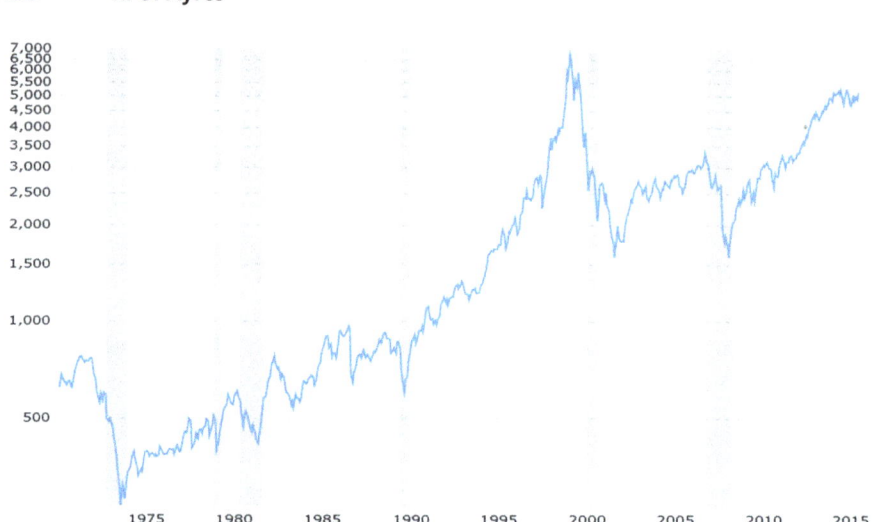

Fig. 5 Nasdaq 100 (1970–2015)

slow. Share values on all indices passed the 2000 peak in 2016 and are now (2017) significantly higher.)

Meanwhile, the people who had bought houses at inflated prices—any time after 2003 or so—were in trouble because house price were dropping as fast as they had risen. This meant that everybody who had a mortgage based on an inflated price lost home equity. A few lucky gamblers who bought on the way up, and resold near the top, made money. This loss hit all but a few homeowners, not just the fraction who bought new houses with "sub-primes." In aggregate, it wiped out about 45% of the home equity portion of middle-class wealth in the USA as of 2006. That was the biggest hit of all on the middle class.

Worse, as of the end of September 2011, 10.7 million homeowners (about 22% of the total) had *negative* equity. This amounted to $700 billion, or about $65,000, per household. It is safe to assume that for most of those people, the loss of home equity wealth exceeded 100% of their life savings. They suddenly found themselves deeply in debt thanks to financial games by mortgage companies and banks, in which they had no part. (My younger brother was one of those. He had invested his total savings in a Seattle Condo, which suddenly became worth less than the mortgage.) National wealth, as household net worth, fell by about $16 trillion dollars from peak to trough (see Fig. 6). Only part of that loss has been recovered since 2009.

In case you happen to think that the age of bubbles is over and that the 2007–8 bubble was the last one, think again. The next graph, Fig. 7, is only

Bubbles and Panics Since 1920   151

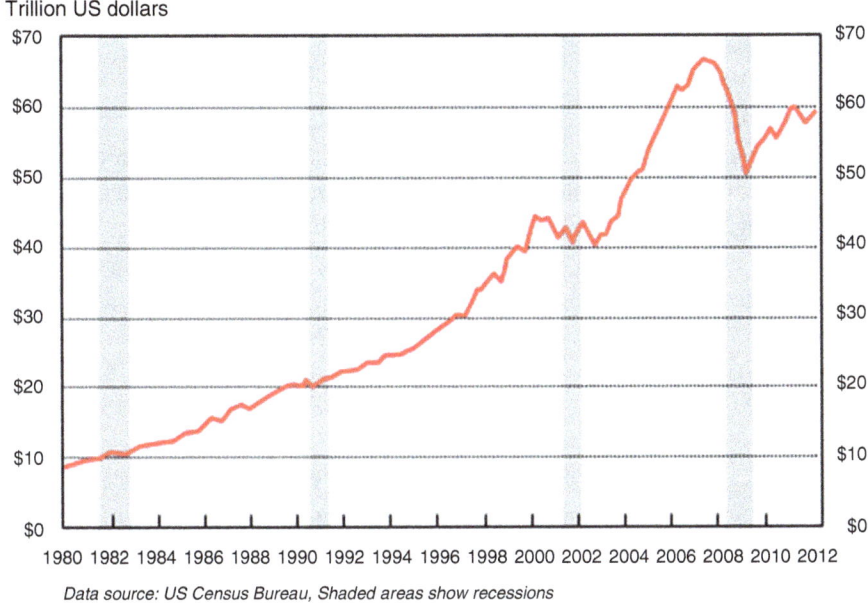

Data source: US Census Bureau, Shaded areas show recessions

Fig. 6   US household net worth 1980–2012

Fig. 7   Rare earth stocks: a classic bubble

the last one I personally got caught in. I really expected a boom in rare earth stocks. There was a boom, but I invested (fortunately not all of my savings) right at the top.

A number of more recent panics and bubbles are discussed in Michael Lewis' fascinating book "Panic: The story of modern financial insanity" (Lewis 2008). This book covers all the major panics on the US stock market from 1987 through 2007. The main message of that book is that panics are much more common than they "should be" according to economic theory. People now talk wisely about "black swan" events (i.e. events that are as unlikely as a black swan) but a lot more devastating (Taleb 2007).

So, dear reader: whenever you see an event, such a hurricane, or a flood, described in the press of the media as occurring "once in a thousand (or million) years," it is important to be aware that any such calculation is based on two underlying assumptions. One assumption is that previous experience will continue to be relevant, i.e. conditions have not changed. The second assumption is that the events in question are "random fluctuations." In mathematical language, this means that the probability distribution of fluctuations, by magnitude, is "normal" (i.e. Gaussian) or (in some cases "log-normal," meaning that the logarithms of the variable are Gaussian). But experience now tells us that both of these assumptions about probabilities are often false, especially as applied to market phenomena.

Since the stock market crash of 2008–09, it has been clear that "normal" probability theory, based on random fluctuations, just does not apply in markets. Humans operating in isolation may make decisions based on purely rational considerations and fuzzy information. In such situations, there may be a "right" or optimal answer, but most people will miss the optimal point by some degree (call it an error) in one direction or the other. In this case, the distribution or errors will be "normal" (Gaussian).

But in crowds, where everybody can see what "the market" is doing, people are notoriously inclined to follow trends. People who are not trained to be skeptical find it all too easy to believe that the growth of a bubble is really a long-term trend. Many people, including professional traders, have adopted theories based on past crowd behavior such as the "Dow Theory" or the "Elliot Wave Theory" or simply "follow the leader" (do what the "sage of Omaha" does) in buying and selling decisions.

The first successful arbitrageur strategy, using computers, was based on "fishing for market inefficiencies" (mis-pricing). It enabled smart computer-literate traders to make a lot of money, starting in the 1960s (Thorp, Kassouf 1967). Thorp and Kassouf's work undoubtedly influenced the work of Fischer Black and Myron Scholes at MIT, a few years later. In 1972, the

"Black-Scholes-Merton" (BSM) option pricing model was published as a book chapter, and a year later in an academic journal (Black, Jensen, & Scholes 1972; Black, & Scholes 1973). Finance professors in business schools quickly adopted it and taught it.

Robert Rubin of Goldman Sachs was a member of the Chicago Options Exchange Board from 1973 on. By 1975, BSM was being used as the method of pricing options (puts, calls, and futures) traded on the floor of the Chicago Board of Options Trading (CBOT). The ban on commodity options was lifted by Congress in 1976. By the early '80's, S&P futures contracts also began trading at the Chicago Mercantile exchange. In 1983, Fischer Black was hired away from MIT to organize the "quantitative strategies group" at Goldman Sachs. (They pay much higher salaries than MIT.)

The first major use of the BSM model by the "quants" was to identify mis-pricing of funds as compared to the component stocks. Then the trader's strategy would be to "short" the overpriced component and "long" the underpriced components. The most famous example of mis-pricing (which did not depend on the BSM model) was the Value-Line futures index. This index was always overpriced as compared to the underlying stocks because it used a geometric average instead of an arithmetic average. This simple discovery by Black (together with the automated direct order trading system initiated by Goldman Sachs) resulted in a $20 million profit for Goldman Sachs with no risk exposure at all (Ellis 2009 #7254, p. 414).

Now, at last, it is time to tell the cautionary tale of Long Term Capital Management (LTCM), one of the first "hedge funds." It was founded in 1995 by John Meriwether, former head of *arbitrage* at Salomon, with some of his "quant" associates, including Robert Merton, Myron Scholes and Jon Corzine, and others. You ask, what is "arbitrage?" Be patient. Arbitrage is what LTCM was created to do, and did. They identified pairs of assets (bonds) for sale in different markets, but with the same expiration date. Both members of the pair should theoretically have had the same value, but for some reason, they differed. Meriwether et al. decided that one of the pair was overpriced and the other was underpriced. They assumed that, in the long run, the two different prices would migrate toward each other and converge, i.e. the overpriced bond would lose value and the underpriced bond would gain value. Their strategy was to "short" the overpriced bond and buy the underpriced bond "long". The difference between the two prices, above the cost of exchange, was pure profit if (and only if) the "true price" was in between the high and the low.

By buying long and selling short simultaneously, the net cost of the transaction was near zero, but the price difference on each transaction was pure profit. Moreover, the quants running LTCM thought that the profit was guaranteed. So they borrowed money from the banks to buy many copies of this short-long combination. That multiplier was the "leverage." In its first year, LTCM achieved an annualized return over 21% p.a. In the second and third years returns hit 43% and 41%, respectively. At that point, the asset value of LTCM had increased from the initial $1 billion to $7 billion, based on an estimated leverage factor of 16, meaning that they borrowed $16 for each $1 of their capital. They thought they had discovered *el dorado*—they could not lose.

At that point, greed stepped in. To make still more money, they needed more leverage. So the partners returned over $2 billion to their (lucky) outside investors in order to increase the fund's leverage to 125:1 (Kirk 2009). It was a mistake. A year later, LTCM collapsed and all of those paper profits disappeared and the founder's capital disappeared as well. The underlying reason was that LTCM had a lot of exposure to Russian bonds, and there was a political crisis in Russia. During that period, holders of Russian bonds panicked and made "irrational" decisions contrary to the logic of LTCM's "sure" bet (Ayres 2014, pp. 89–90). By a strange twist of fate, Scholes and Merton received the Nobel Prize in economics in 1999, just a year after the collapse of LCTM.

The worm turned in 1998. Following the 1997 Asian financial crisis and the 1998 Russian financial crisis, in 1998 LTCM lost $4.6 billion in less than 4 months. At that point, the Federal Reserve had to step in, although doing so is not part of the Fed's mandate. Under the Fed's stewardship, an agreement by 16 financial institutions resulted in a $3.6 billion recapitalization bailout. The founders were also out and the hedge fund was liquidated and dissolved in 2000. The technical reasons for the failure of the arbitrage strategy in that (and other) cases is interesting in itself, but not relevant here. What happened in 2008 was that, in times of panic there can be irrational behavior in the markets. In that case, the assumption of "normal" distribution of random fluctuations is no longer valid. Fluctuations may be engineered by other players in the game. Other players may notice that "someone" is betting consistently against a trend, and that the bets are increasing in size. This situation can resemble a game of poker, in which a player with a weak hand is bluffing and betting heavily to drive the other players out of the game. The poker game can become a contest between two players where the one with the deeper pocket can sometimes win just by betting heavily. There are indications, for instance, that the very sharp and sudden decline in oil prices in 2014 was caused the actions of one player with very deep pockets (Saudi Arabia) that wanted to drive other players (e.g. shale "frackers") out of the game. It was the

United Copper squeeze play all over again, and it also failed, luckily without the after-effects.

Another academic contribution to the theoretical confusion is the "Efficient Market Hypothesis" (EMH), introduced half a century ago by Eugene Fama, then a graduate student (now a professor) at the University of Chicago (Fama 1970). His hypothesis was that markets are "informationally efficient" in the sense that at any moment they reflect all the available information that is pertinent. The main implication of the EMH hypothesis is that current market prices do not depend on history, only on current information. If this is true, then nobody can forecast future stock market prices. If that is true, it follows that successful investors just happen to be the luckiest ones.

According to the EMH, there is no such thing as a "bubble" in the stock market because markets always reflect the total of all available information and therefore prices are always "correct," even from moment to moment. To put it simply, EMH does not allow for market irrationality. (Fama received the Nobel Prize for economics in 2014.) One weakness of EMH is that it effectively assumes that information is propagated at infinite velocity. This is clearly not true, and many effective trading strategies, notably High Frequency Trading (HFT), depend on getting orders to buy or sell to the front of a queue. (See "The Flash Boys".)

It is interesting that Warren Buffet, the world's most successful "value investor," took issue with Eugene Fama (and EMH) in a 1984 article for *Hermes*, the magazine of Columbia Business School. The article was entitled "The Super-investors of Graham-and-Doddsville" (Buffet 1984). In the article, he named a number of investors, apart from himself, all from the value-investing camp, who had, in fact, beaten "the market" by large margins. Fama tried to explain this as luck. He said that, given a large number of bettors and two choices say, red or black, there will always be a small number who win many times in succession (as well as a similar number who lose as frequently as the winners win.) He argued that Buffet was just one of the lucky ones. Buffet disagreed with this explanation. After all, he is not playing with dice.

A recent book by Pablo Triana, a former derivatives trader who is also an academic, entitled *Lecturing Birds on Flying* makes similar points about models. His target is mathematical models of human behavior in markets that depend too much on notions borrowed from physics (e.g. "Brownian motion"). I can agree with his summary comment: "*Make no mistake, quantitative finance had a very large hand in what could well be the worst financial crisis in the history of mankind*" (Triana 2012). That much is true; it was the use, and misuse, of mathematical models, especially BSM and another one called Value at Risk (VaR), that encouraged hedge funds, in particular, to increase their leverage so much in 2005–07, and get burned.

To summarize, there was a time, roughly from 1985 to 2005, when the smartest hedge fund managers made money using algorithms that identified market inefficiencies of the sort discovered by Thorp and Kassouf back in the '60's. It is also increasingly clear that the "mis-pricing" opportunities they found are getting scarcer and that many "insurance" strategies such as "portfolio insurance" do not work in panics.

A new trend in the markets, mentioned above, is known as "high frequency trading" (HFT) by ultra-fast computers. These "robot" traders operate independently of human intervention or judgment. IBM and Hewlett-Packard independently developed such algorithms in 1996–97 (MGD and ZOP, respectively). In 2001, an IBM group showed that HFT outperformed human traders and estimated the overall benefits (to broker-dealers) at several billions of dollars per year. That triggered a race. Deutsche Bank developed its "Stealth" algorithm. Credit Suisse now uses "Sniper" and "Guerilla." The champion of all the hedge funds using HFT is James Simon's Renaissance Technologies, which develops its own algorithms and does it better than anybody else.

By 2006, a third of all market transactions in the USA and Europe were done by HFTs. By 2009, the HFT share in the USA was up to 79%, and it seems likely that the current share is even higher. In fact, HFT is now *de rigueur* for all major financial institutions on both the buy side and the sell side. This fact is very bad news for individual traders or small brokerage firms who cannot compete.

The benefits of HFT overall—if any—cannot be easily measured. But there are several market strategies that utilize HFT, especially arbitrage. (The word is jargon for tracking prices for the same security in several different markets and profiting from differences, by buying one long and shorting the other.) This was exactly the original LTCM strategy. Evidently prices can change quickly, so it becomes important to carry out both sides of these transactions simultaneously so as to minimize the potential for loss due to price changes during the order execution (known as "execution risk").

But robot traders are fundamentally stupid: they do only what they are told to do in specific circumstances. On May 6, 2010, an event occurred that demonstrated this point dramatically. A major trader (alleged to be Kansas-based Waddell and Reed, a mutual fund) decided to sell a large number ($4.1 billion worth) of E-mini futures contracts for the Standard and Poor 500 index fund. The sale was executed by an algorithm-driven robot. What happened? The Dow Jones Industrial Average (DJIA) dropped 481 points in 6 min and then recovered by 502 points 10 min later. From an overall market-watcher's perspective, it was a "blip" of no particular significance. However, as the later report by the SEC and the commodity futures trading commission (CFTC) noted, "over 20,000 trades across more than 300 securities were executed at prices more than 60% away from their values just moments before" [http://www.market-watch.com/story/text of flash crash—reports its summary]. In fact, during one period of just 14 s, 27,000 automated orders were executed between HFTs, but only 200 E-minis were bought "to keep," so to speak.

In essence, the HFT's were selling to each other during those seconds, because there was not enough cash (liquidity) in the system to absorb all the sell orders. So, robot traders were buying E-minis because they were programmed to do so, whenever the price fell below a certain level. But every time such a purchase brought the allowed inventory for that fund above a preset level, the robot trader proceeded to re-sell them to other robot traders that were programmed to buy that security. The ownership of those orphan E-minis was passed round and round the merry-go-round, until enough liquidity (cash) became available. This behavior became known as the "hot potato effect." Luckily it did not do much damage, on that particular occasion.

But more important, there were other un-anticipated chain effects. For instance, an automated purchase of one security (such as an E-mini) would force a buyer to sell some another security, to maintain the cash value of the fund. Of course, this could (and did) result in a chain of still other forced transactions. It was the chain effect that pushed the DJIA down so fast. Luckily, the market re-equilibrated in a short time, and the DJIA rose back up to its earlier level. In this case, the faulty assumption behind the "flash crash" was that there would always be enough liquidity in the system to absorb the securities being offered for sale.

Regulators are still wondering whether this kind of (unintended) behavior constitutes a major hazard to the markets and, if so, what to do about it. An obvious remedy would be to put an upper limit on the size of the sell or buy order that can be executed at one time, in relation to the total trading activity in that market over some previous time period measured in hours or days, not milliseconds.

There were always losers as well as winners, but in the five golden years 2003–2007, the winners using models won a great deal of money from a large number of losers, who hardly noticed because each loss was so small. Naturally those winners attracted a lot of savvy investors: assets under management tripled from about $600 billion to $1.8 trillion. The hedge fund industry (as of 2017) had over $3.15 trillion in assets under management. Yet the average hedge fund lost money in 2011 and even the top 10% of funds only made returns of 19.5%, far less than they had made in their heyday.

That performance by the top tier—quite good by some criteria—was less than half of the returns made by the top 10% of hedge funds in every single year from 2000 through 2010 [Dan McCrum, FT Sept. 11, 2012]. McCrum attributed this decline to the fact that statistical information from Bloomberg and other sources is now so easy to tap into that the number of money-making opportunities from "inefficiency-fishing" has declined significantly. He calls it *"the end of alpha,"* meaning that it is getting much harder to beat the market using computers. Let us hope that the plot of the novel (and movie) "The Fear Index" by Robert Harris was just an imaginative fiction.[5]

One change in the historical pattern may be the increasing importance of technological bubbles. The "dot.com" bubble of 1999–2000 was not the first bubble related to new "dot.com" technologies. The tulip mania of 1636–37 was based on new varieties produced by growers. The canal boom of 1790–93 in England was less irrational, but still a bubble. The railroad boom of 1845–47, also in England, was another example based on technological change. So was the Western Pennsylvania oil boom of 1859–66. There have been many books (and busts) associated with resource discoveries, especially the California gold boom of 1849, followed by the Klondike gold rush of 1896–99 (and several in South Africa), the Florida land-boom of the early 1920s and the stock market boom (led by radio) of the late 1920s.

The "bitcoin bubble"—allegedly driven by a new computer technology called "block-chain"—appears to be another example of the bubble phenomenon. A bitcoin that was worth $8 at the end of 2016 reached $16,000 recently although it has fallen quite a bit since then. Like most bubbles, this will crash. In this case, the crash is inevitable because the cost of "block-chain" computation required for any bitcoin transaction makes it uneconomic, except for extremely large transactions. Crypto currencies offer very little practical value for paying bills to anyone, except to multinational companies and criminals. An unregulated currency would be very dangerous, in any case.

There is an analogy between financial market collapses and avalanches or forest fires. In terms of the avalanche analogy, these headwinds can be regarded cumulatively as the weight of the snowfield on a steep mountain slope. Avalanches are a regular feature of mountain terrain. They cannot be avoided completely. But their magnitude (and resulting damage) can be reduced, mainly by deliberately triggering small and frequent ones to prevent the snow buildup from getting too big.

---

[5] The title of Robert Harris's thriller, "The Fear Index," comes from the volatility index, or VIX—also known as the "fear index"—which measures expectations of violent swings in the market. The plot is a variation on Mary Shelley's "Frankenstein" viz. a scientist who creates an intelligent computer program named VIXAL that evolves into a form of artificial intelligence.

A similar mechanism can be seen in forest fires. During periods of drought, the weaker trees and shrubs dry out and the forest biomass—living and dead—becomes increasingly inflammable. Lightning strikes from electrical storms routinely cause small fires that usually burn themselves out locally. But when and if the forest gets drier, the small fires get hotter and they spread faster. At some point, big fires generate their own cyclonic windstorms that can carry sparks and embers considerable distances and make the fire virtually uncontrollable. This is what happened in the record-breaking 2017 fire-season in San California, when over 9,000 fires burned 5,590 km^2, also causing enormous economic damage. As of mid-2018, the fires were even larger. Again, forest fires cannot be eliminated completely, but their damage can be minimized by not allowing a buildup of inflammable material on the ground.

We need to find equivalent strategies to contain potential financial firestorms. A more important problem is to use money to settle conflicts wherever possible, even where the issues are not fundamentally economic.

Economic Cycles, in Principle

Cyclic booms and busts have been a feature of economic life—and one of the major arguments against capitalism—since the eighteenth century. The tendency for "Boom and Bust" cyclicity has been noticeable since the beginning of the nineteenth century. In fact, it has been a continuing complaint about the nature of capitalism. Karl Marx was arguably the first to offer a semi-coherent explanation of crisis. Here is a quote that illustrates my difficulty with the subject.

> Because crisis is the concentrated explosion of all of the contradictions of capitalistic production, to concretely grasp crisis as such, one must first unfold all of the contradictions of capitalistic production according to their internal relations, and thus the significance of each contradiction as a moment within the totality must be elucidated, and then next one must clarify what processes they pass through and in what sense they must explode in a concentrated manner. And this is above all precisely what Marx sought to achieve throughout his entire critique of political economy (of which *Capital* is the most basic part). In other words, *Capital* at the same time can be said to include a theory of crisis (its most fundamental part). By Samezo Kuruma, An overview of Marx's Theory of Crisis. Japanese title: Marukusu no kyōkō-ron tenkiyō;
>
> First published: August 1936 issue of *Journal of the Ohara Institute for Social Research* Source: Chap. 3 of *Kyōkō kenkū* (Investigation of Crisis), Tokyo: Otsuki Shoten, 1965;
>
> Translated: for marxists.org by Michael Schauerte; CopyLeft: Creative Commons (Attribute & ShareAlike) marxists.org 2007.

In simpler terms (as I interpret this), Marx's *Das Kapital* attributed cycles to a contest between overproduction resulting in declining returns to scale (a standard assumption of most economic theory) and under-consumption due to falling wages as producers cut costs in order to cut prices (Marx 1867). Marx focuses more on the "contradictions" that lead to collapse than on the mechanisms that lead to recovery.

During the great depression, Arnold Schumpeter took up the subject and wrote *Business Cycles: A Theoretical, Historical and Statistical Analysis of the Capitalist Process* (Schumpeter 1939 #4532). Since Schumpeter's typology, the main economic cycles are nowadays named after their discoverers or proposers (Schumpeter 1939). They include the following:

- 3- to 5-year inventory cycle, identified by Joseph Kitchin (1923);
- 7- to 11-year "fixed investment" cycle, identified by Clement Juglar (1862);
- 15- to 25-year infrastructural investment cycle, identified by Simon Kuznets;
- 45- to 60-year long-wave technological cycle of Nikolai Kondratieff.

These are the best-known, but a few others have been proposed, including a building cycle, suggested by Arthur Burns, and the theoretical Goodwin and Minsky cycles to be discussed later. Kitchin studied the existence of "Minor Cycles" with an average length of about $3\frac{1}{2}$ years or 40 months, for which he concluded that two or three "minor cycles" fit within one Juglar (or Business) cycle. Schumpeter described the Great Depression as the simultaneous convergence of the Kitchen, Juglar, and Kuznets cycles.

J.S. Mill called attention to a land-price speculation cycle in his "Principles of Political Economy" (Mill 1848a, 1848b #5992) Book III, Chap. XII. Henry George was strongly influenced by the recession of 1873. He identified land-price speculation around cities as the primary cause. The mechanism—with-holding land from development until the value has risen—followed by overbuilding with borrowed money, is described in his book (George 1879). Since then, a variety of studies have identified an 18-year land-price cycle (Harrison 1983), Harrison 2005). Based on this research, Fred Harrison correctly (but retrospectively) predicted both the 2001 and 2008 financial crashes.

The inventory cycle is conventionally explained as a buildup of stocks in the inventory of retail shops, resulting in a downturn in orders to manufacturers, and a slowdown in production. When the inventory oversupply is used up, orders increase and production increases again. The GDP may not actually turn negative during the inventory disinvestment phase, but the rate of

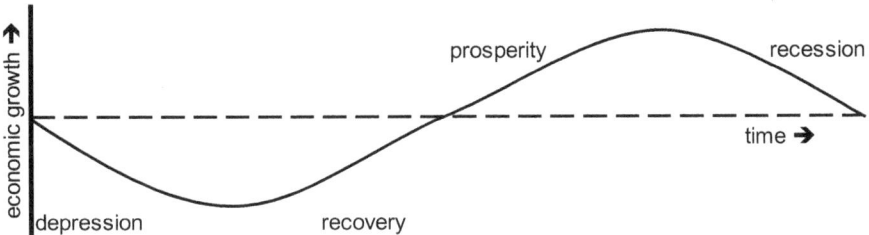

Fig. 1 The Schumpeterian wave

overall growth decreases. The period of the cycle is roughly the time it takes to increase and decrease production and/or the time it takes for consumers to reach the limits of their ability to buy.

The first conceptual breakthrough insight (apart from that of Marx) came from Juglar (1862), who shifted focus, no longer looking at the isolated problem of the individual crisis. He described business cycles as wave motions in which crises followed periods of unsustainable growth and overspeculation. Schumpeter contributed to the development of the cyclical theory by developing the asymmetric three-phase model of Juglar (1862) into a sinusoidal four-phase model consisting of prosperity, recession, depression, and recovery (ibid, p. 167), of which the depression phase is not necessarily included in the cycle. The Schumpeterian wave is indicated in Fig. 1.

Certainly, the Juglar "cycle" of 7–11 years seems to be correlated closely with successive economic crises, not waves. Kuznets (1930) investigated second-order secular wholesale price trends and concluded that they are related to demographic phenomena and immigration. He also claimed to find a relationship between his swings and income inequality, with inequality rising during booms and declining during recessions. His results were sharply criticized at the time, and seem less relevant today than they did in 1930. The subject became central to macroeconomics after the National Bureau of Economic Research took it up in the 1940s. That research program culminated in the publication of the book "Measuring Business Cycles" by Arthur Burns and Wesley Mitchell (Burns & Mitchell 1946).

The possible existence of "long cycles" in prices and economic activity about 50 years from peak to peak was noted more than a century ago by Jevens (1884) cited by (Kleinknecht 1987). In fact, Jevons cited even earlier articles. However, the first author to subject the hypothesis of long cycles to systematic analysis was the Marxist Dutch economist, Jacob Van Gelderen (Van Gelderen 1913), who anticipated much that has later been "rediscovered" by others.

The existence of long-term economic waves has been widely acknowledged after the 1925 publication of the Soviet economist, Nikolai Kondratieff, in the "Voprosy konyunktury," on "Long Economic Cycles" (Tarascio 1988). A German translation of Kondratieff's article was published in the "Archive für Sozialwissenschaft und Sozialpolitik" (1926) and an English summary of the German article was published by Kondratieff (1926).

Kondratieff (1926) covered a period of 2½ waves, of which the first two were 60 and 48 years. Based on this analysis, he indicated an average length of about 50–55 years. Many others followed him in his research. In the absence of a convincing cyclic theory, the Kondratieff cycles are generally referred to as K-waves. Table 1 provides an overview of the various Kondratieff waves that have existed since the beginning of the Industrial Revolution.

Van Gelderen (1913) was the first to suggest a plausible causal hypothesis, even before the waves were discovered. He suggested that a long period of rising prices (prosperity) is driven by the rapid growth of one or more "leading sectors." Van Gelderen also discussed and tried to explain other important features of the process, including periodic over- and under-investment of capital, periodic scarcity and abundance of basic resources, credit expansion and contraction, and so on. Finance is obviously relevant.

Schumpeter's well-known study of business cycles was, in many ways, an extension and update of Van Gelderen's ideas (Schumpeter 1939). He proposed that *temporal clustering* of a number of major technological innovations during periods of deflation and recession might account for the dramatic growth of the so-called leading sectors which seems to drive the inflationary half of the cycle. This idea was immediately and sharply challenged by Kuznets, who doubted both the existence of Kondratieff cycles, and the causal explanation suggested by Schumpeter. However, Kuznetz seems to have taken the idea rather more seriously in a later book.

Table 1 Historical overview of past Kondratieff waves

Kondratieff	1st wave	2nd wave	3rd wave	4th wave	5th wave
Depression	1764–1773[a]	1825–1836	1872–1883	1929–1937	1973–1980
Recovery	1773–1782[a]	1836–1845	1883–1892	1937–1948	1980–1992
Prosperity	1782–1792	1845–1856	1892–1903	1948–1957	1992–2000
Prosperity	1792–1802	1856–1866	1903–1913	1957–1966	2000-2009[a]
(War)	(1802–1815)		(1913–1920)		
Recession	1815–1825	1866–1872	1920–1929	1966–1973	2009–2018[a]
Length	61 years	47 years	57 years	44 years	45 years
Excl. War	48 years	47 years	50 years	44 years	45 years

Source: Personal Correspondence, originally based on van Duijn (2007)
[a]Estimated on the basis of an average 9-year Juglar Cycle.

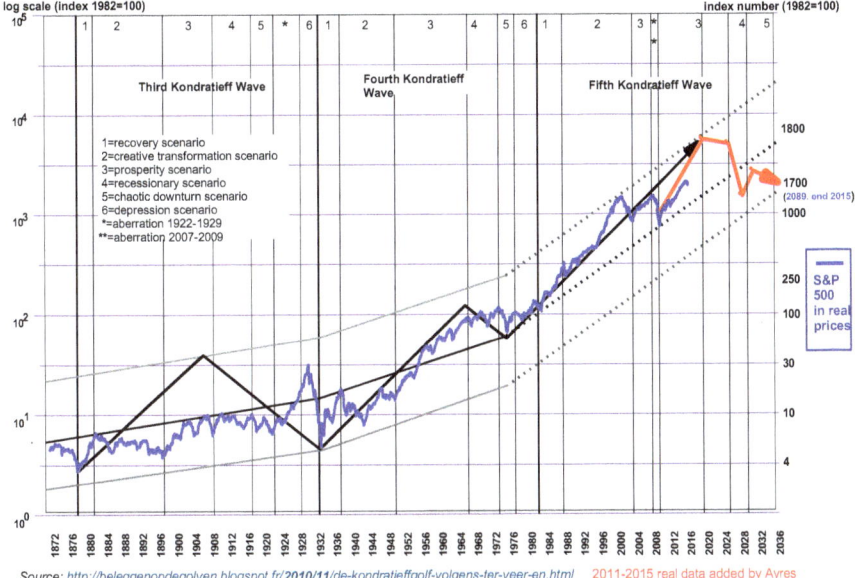

Fig. 2 Kondratieff chart expanded to 2035. The red part of the curve, added by RUA, is obviously guesswork but has some basis in theory

The graph (Fig. 2) depicts the Kondratieff cycle in historical context, extrapolated into the future. The "third wave" (1876–1932) is far from obvious though the fourth and fifth waves, ending in 2008, seem evident. A sixth wave starting in 2009 (in red) is pure conjecture. David Kotz conjectures that the decline of the long wave after 2000 is due to the collapse of neo-liberalism.

The subject of long waves has been revived yet again since Schumpeter's work, especially by W.W. Rostow (Rostow 1975, 1978), Gerhard Mensch (Mensch 1979), and Jay Forrester (Forrester, Graham, Senge, & Sherman 1985). For more recent updates, see Freeman (Freeman 1996). Rostow's interest was primarily directed to the phenomenon of "takeoff", leading to sustained long-term economic development. He viewed Van Gelderen's leading sectors' as not only the drivers of the long wave, but as the engine of long-term growth for the whole economy. Mensch attempted to explain the gaps between innovation clusters by invoking a theory of investment behavior. He postulated that during periods of general prosperity, investors will shy away from risky long-term ventures (innovations), whereas during periods of stagnation or recession they may be more willing to invest in new ventures. Forrester was attempting an integrative approach.

In this context, the investment-debt cycle considered by Richard Goodwin appeared relevant, even though it is not really part of the long-wave literature (Goodwin 1967, 1987). Goodwin's model exhibits a cyclic behavior of borrowing and repayment to banks for purposes of investment in capital stock and growth. At first, when debt is low, the growth is brisk. But as demand approaches capacity, costs increase (and wages rise) so profits decline. Meanwhile, there is an accompanying buildup of debt to the bank. This leads to an eventual collapse of the economic system as debt service consumes more and more of the economic surplus from the non-bank sectors.

The collapse ends with a large debt write-off and the cycle begins again. This usually accompanies a war, a major recession or the bursting of a large financial bubble. It is worthy of note that the Goodwin bubble mechanism works (theoretically) in a "normal" economy, meaning one driven only by the accumulation of capital stock per worker, without any Schumpeterian "creative destruction." The Goodwin model does not explain real financial bubbles, but it has spawned more realistic models that will be discussed later in connection with models of disequilibrium.

At least one important variant of the Mensch thesis, associated primarily with Freeman and his coworkers, has emerged from this debate (Dosi, Freeman, Nelson, Silverberg, & Soete 1988; Freeman 1996). It is that the rapid growth period of the long wave is not necessarily driven by innovations occurring in the immediately preceding "trough." There seem to be other cases where the rapid growth period was driven partly, or mainly, by the adaption/diffusion of important technologies that were tentatively introduced much earlier, but which needed a long gestation or were not yet "ripe" for some reason.

This notion does not dispute the importance of the basic innovation (or the key facilitating inventions preceding it), but it does put major emphasis on the subsequent processes of development, improvement, application to new (and sometimes unexpected) purposes, and subsequent adoption. In all this, there is a continuous and vital feedback between the innovator and the user, characterized by learning on both sides. The technology diffusion process as this set of interactive phenomena is usually called, thus becomes quite central to any complete theory of long waves.

The latter thesis, in turn, has spawned a more formal theory of technical change and innovation/diffusion in economics, beginning around 1980. The major conclusion of this body of work is hard to summarize, beyond saying that innovations do not come out of thin air. They take place when the underlying technical capabilities are present, financial support is available from government or entrepreneurs, and they are needed to solve a pressing social or economic problem. But when the conditions (including the last) are met, inventions sometimes occur simultaneously in different places. (The telephone, the carbon-filament light-bulb, and the Hall-Heroult aluminum smelting process are the classic examples of simultaneous invention.)

Globalization and the Decline of the Labor Movement

In the real world, the employers ("job creators") and their employees must exist in a balance. Each depends on the other. Job creators (i.e. organizations that produce things and services) need workers and customers (consumers), and consumers need jobs. Henry Ford paid his workers more than the national average, back at the beginning of the twentieth century, with the long-term objective of making his workers into his customers. Economic growth in the 1950's and '60's maintained that balance by splitting productivity gains between capital and labor.

But in recent decades, especially after 1980, virtually all of the productivity gains have gone to the owners (shareholders) and their agents, the top executives, and almost none has gone to the employees. Henry Ford's innovation, making his workers into customers, has been undone and reversed. People belonging to the census category "employees"—persons paid for their work—received between 52.5% of GDP and 57.5% of GDP between 1950 and 2015, in the USA. This percentage peaked in 1970. But the subcategory of employees receiving wages or salaries paid monthly was 49% in 1950, but has declined to 43% in 2015. The difference consists of people with part-time jobs or temporary "gigs" (see Fig. 1 below). The percentage of wage earners belonging to a union has declined even further.

Within the top 1%, a small fraction (the top hundredth of a percent) have captured most of the gains. How and why has this happened? A major part of the explanation can be attributed to the leveraged buyout (LBO) movement and the shareholder value maximization (SVM) movement as discussed in a previous chapter. But there is quite a bit more to be said.

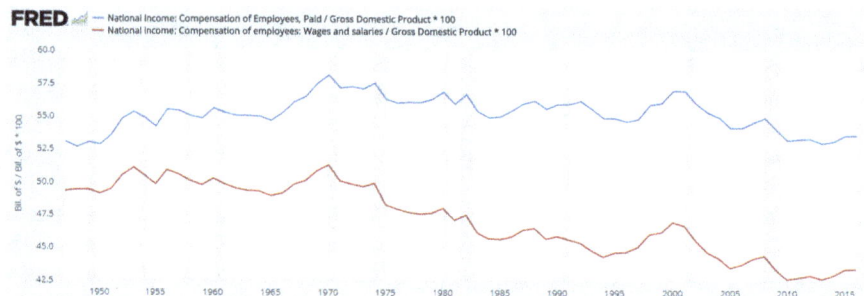

Fig. 1 US employee compensation as a percent of GDP, 1950–2015

One is that changes in the law have weakened the bargaining position of industrial workers. When the labor movement began in the nineteenth century, it was a contest between owners of a local factory, railroad or mine and the local workers. In principle, the workers stopped work if no agreement could be reached. This rarely sufficed because it was too easy for employers to hire "strike-breakers" in most industries. Organizations like the "detective" agency, Pinkertons, grew up to furnish strike-breaking services on demand.

Then the unions tried to get the law changed to prevent the owners from hiring strike-breakers. The employers did not only continue to hire strike-breakers, they also often hired goons to beat up and discourage union organizers. Industrial unions typically had to save money for several years in order to finance a strike. When the strike hurts only the strikers themselves and their local employer (who can lose profits and market share), it is a bargaining situation comparable to any free market bargain between buyers and sellers, even if the pain is unequally divided between the parties. Third parties were not much affected, beyond minor inconvenience. The American Labor Movement peaked in the depression years. Its greatest triumph was the unionization of GM by the UW in 1937.

But in the postwar years since 1947, "free trade" agreements were negotiated around the world, justified by a trade doctrine mainly associated with the names of Adam Smith and David Ricardo, at the beginning of the nineteenth century. At the time, "free trade" was opposed to "protectionism" (by high tariffs). The economic argument for removal of trade barriers was, simply, that it would enable the most efficient producers to grow, by exploiting economies of scale, resulting in lower prices for consumers and economic growth. (The "wealth of nations" was all about trade.) Political support for free trade in England was primarily from urban consumers who benefited from lower prices of grain ("corn"), and exporters of manufactured textiles. Opposition came from landowners and farmers afraid of competition from the American Colonies.

The first victory of free trade ideology over protectionism was marked by the repeal, in 1846, of the English "corn laws" (tariffs on imported grain, introduced in 1815). This political earthquake was partly inspired by the Irish potato famine (1845–49). That famine was due to a fungal disease, but it was amplified by monocrop agriculture and overproduction for purposes of export. It led to lower prices for basic foodstuffs for the urban population of England and starvation for the Irish farmers. Thereafter, England became the global champion of free trade. It is noteworthy that *The Economist* newspaper (actually a magazine) was founded in the nineteenth century to promote free trade. This publication reiterates its editorial message re-trade in virtually every issue to the present day. In the nineteenth century, foreign investment (and repatriation of profits) moved some jobs away from the farms, but rapid growth created new ones. Moreover, emigration to the Americas was an option for virtually anyone. That option is no longer so readily available.

Free trade agreements have multiplied since the creation of the General Agreement on Tariffs and Trade (GATT) in 1947 (now the World Trade Organization (WTO), since 1995). These agreements have enabled manufactured goods and capital (but not workers) to move more freely around the world. The considerable reduction in "protectionism"—meaning protection of producers and their wage earners in one country—has allowed many big businesses to move manufacturing activities to low-wage countries. It has also had the effect of suppressing wages in wealthy countries. This has pushed prices of imported goods down. There are some benefits to "consumers" of the imported goods that are no longer produced locally. But the export of jobs (in the US case) has resulted in a negative balance of trade. That, in turn, tends to slow economic growth. Job export has also "hollowed out" the middle class by forcing domestic workers in the rich countries to compete against foreign workers in countries with no limits, to the use of child labor, long working hours, and lack of safety and environmental standards. It has become a "race to the bottom."

Globalization has also created jobs in the low-wage countries, especially China, pushing their wages up and increasing their consumption. Eventually some of that increased consumption becomes demand for US products (such as i-Phones). But the i-Phones are now also made in China. In short, imports by the low-wage countries from the wealthy countries do not necessarily compensate for the lost income by the displaced workers in industries that have moved. Since the 1950's and 1960's, most of the TV industry, the bicycle industry, the watch industry, the sewing machine industry, and much of the machine tool industry and the auto industry have moved out of the USA. Even the "hardware" part of the consumer electronics industry is now mostly located in South Korea, Taiwan, China, or Malaysia.

This has enabled Japan, Korea, and China to industrialize rapidly, but it has made "vagabonds" of the formerly well-paid ladies garment workers in New York City, and the automobile factory workers in Detroit or Birmingham. Today, industrial employers simply move their factories to low-wage countries, where they can make their goods without unions to bargain with. (They bargain instead with the local power brokers). "Free trade" agreements enable them to retrieve their profits (or move them elsewhere to escape the tax-man). It is the other reason why wages have been static or declining in the USA for the past 40 years.

Thanks largely to this global relocation phenomenon, manufacturing (in the USA and Europe, including Germany) has been steadily declining as a source of value added. More and more jobs are in the service sector, or the construction sector, and most of those jobs cannot be moved easily to other countries. On the other hand, a lot of service jobs are repetitive and can be automated. Friendly elevator operators and telephone switchboard operators are things of the past. Bank tellers have been replaced by ATM machines, and telephone calls are already largely automated, saving money for employers, but imposing significant time costs on callers. Other employment categories, such as drivers, clerks, or waiters, are becoming candidates for robots.

Globalization has also triggered another phenomenon. Strikes by some industrial workers (e.g. truck drivers, longshoremen, railway employees, or airline pilots) and many public sector workers (e.g. teachers, nurses, police, or air traffic controllers) hurt the public more than the "owners" or the administrators of the schools, hospitals, or towns. This gives some groups of workers exceptional bargaining power, in proportion to the amount of harm to the public a strike can do. In some instances, it has resulted in large wage gains or pension gains by teachers, firemen, or police, quite unrelated to productivity.

In such cases, the consumers of public services and the taxpayers become pawns. In economic language, the third-party effects are "externalities."[1] Understandably, if irrationally, the public response to this externality has been increasingly anti-union. In the USA, it has enabled legislators, egged on by employers' lobbies, to pass "right to work" laws that deliberately make it hard for unions to organize the service sector. The net result has been to allow employers in some industries (such as garment workers) to move to "non-union" environments where wages can be kept low because neither individual

---

[1] N.B. when welfare economists first noticed externalities in the 1920s, and thought that they were minor imperfections in the market economy, the suggested remedy was a "Pigouvian tax" on the beneficiaries of transactions causing the externality (e.g. pollution) to compensate the third-party losers (e.g. the public). No such tax has ever been implemented, and it is very hard to see how it could work, even in principle, in the case of strikes against the public sector.

workers nor small business owners have any bargaining power against large enterprises. Walmart illustrates the point. Walmart builds large stores on cheap land outside the centers of towns, pays near—or even below—minimum wages, imports most of its goods from low-wage countries, and attracts customers away from downtown stores thanks to its very low prices.

The Walton's successful strategy, based on "free trade" ideology, has driven literally hundreds of thousands of small "mom and pop" shops out of business, and made the Walton family (Walmart owners) obscenely wealthy. Three of the ten richest people in the USA today are children of the founder of that predatory chain, and they have done literally nothing to deserve their inherited wealth. They do not even create good jobs for the underclass. The Waltons are not alone, of course. Other chain stores that inhabit thousands of malls outside of central cities around the USA were doing exactly the same thing, until Amazon came along. Downtown stores are increasingly not in the business of selling ordinary household goods and groceries to ordinary citizens. Even the malls are closing. (There are very few small owner-manager shops in malls.) Up until now, the labor movement has not found any viable way to counter this trend.

There is an ideology, widely accepted among business, financial, and political elites, that debt repayment is not only a moral necessity but also good for the economy. After WW I, the US government tried hard for years to recover the loans that private banks had made (with government encouragement) to Britain, France, Italy, and the other allies, to fight the First World War in 1914–18. (Not much was ever repaid.) The United States also made large sales of gold to Germany after WWI in exchange for Deutschmarks that soon became worthless. After the Wall Street Crash in 1929, many local and regional banks in the USA were in trouble because their clients (such as farmers and shopkeepers) had no money. But the Federal Reserve, back then, advised by Andrew Mellon and his ilk, raised interest rates when they should have lowered them. That decision made the financial crash into the Great Depression.

Yet austerity is still presented by its advocates (such as German conservative politicians) in terms of "cutting out the rot," as a surgeon might have to cut off a gangrenous leg. The alternative to austerity is painted as collapse of the currency and hyper-inflation, modeled on the German hyper-inflation of 1921–22. The theory for dealing with financial crises is known as the "Washington Consensus." It says that when a government cannot pay its debts (for whatever reasons), cutting government expenditure to reduce debt is the magic key to recovery. This cocktail of measures includes tax cuts for business (to attract investors), privatization of government services and "reform" of labor laws to increase "competitiveness" (i.e. reduce union power). This theory is the justification for the International Monetary Fund (IMF).

As applied to some developing economies in Southeast Asia and Latin America on a borrowing binge, the IMF has had some successes. But as applied to Europe, it is illogical on its face and has utterly failed in practice. What happens, in practice, is that the debt is "bailed out" by loans from the IMF or the European Central Bank, which are just used to keep paying interest on the earlier loans and thus to prevent "infection" of the whole international banking system. The promised inflation has not occurred. Instead, Europe got a taste of deflation. The debt, itself, just keeps increasing although much of it now "receives" (not the right word) negative interest in Germany.

Yet domestic political support in Germany, the Netherlands, and Scandinavia for austerity imposed on the southern countries of Europe remains virtually unabated. These are countries with positive trade balances. The fact that West Germany accepted higher taxes for many years after reunification in 1989—to bail out East Germany—is conveniently forgotten. Who are the winners and losers from policies of austerity? There are no winners in the long run. The austerity hawks in the north will lose their southern markets. The losers are those at the bottom of the income pyramid, especially the young. Austerity as a policy shifts wealth preferentially away from the lower parts of the pyramid. Everyone loses, but the wealthy at the top lose less than the poor.

The decline in the labor movement and the rise in the Walmart business model (and its imitators) have been a major cause—probably the major single cause—of increasing inequality between the bosses and the grunts. I have more to say on this in the next chapter.

Mechanisms for Privatizing Profits and Socializing Losses

In earlier chapters of this book, I have noted several times that the benefits of technological innovations and economic growth are being captured by a small—and ever smaller—fraction of the population whose actual contribution to the creation of wealth is virtually nil. This situation is closely analogous to that of the urban landowners 200 years ago. The landowners profited from the activities of others driving the growth of cities, and did little or nothing to make those activities more productive.

As Richard Werner has pointed out succinctly, profits can be used for just three purposes: investment, consumption, or speculation (Werner, 2005 #7140, 2014 #8424). Speculation is a fancy name for gambling. The activities of bankers and fund managers since the 1970s and especially since the end of the Glass–Steagall era are increasingly devoted to speculation, often with other people's money. For instance, there is a category of financial assets that plays a role comparable to undeveloped land, i.e. "margin loans" collateralized by stocks and shares. In principle, loans collateralized by other financial assets can be used productively, to finance building or even product development. But mostly they are used for gambling in the stock market through investment funds based on empirical theories about market behavior, implemented by computer-based algorithms (see Fig. 1). Sharp rises in market values are closely correlated with margin loans as the graphs show.

This chapter is essentially a partial list of mechanism by means of which wealth is concentrated.

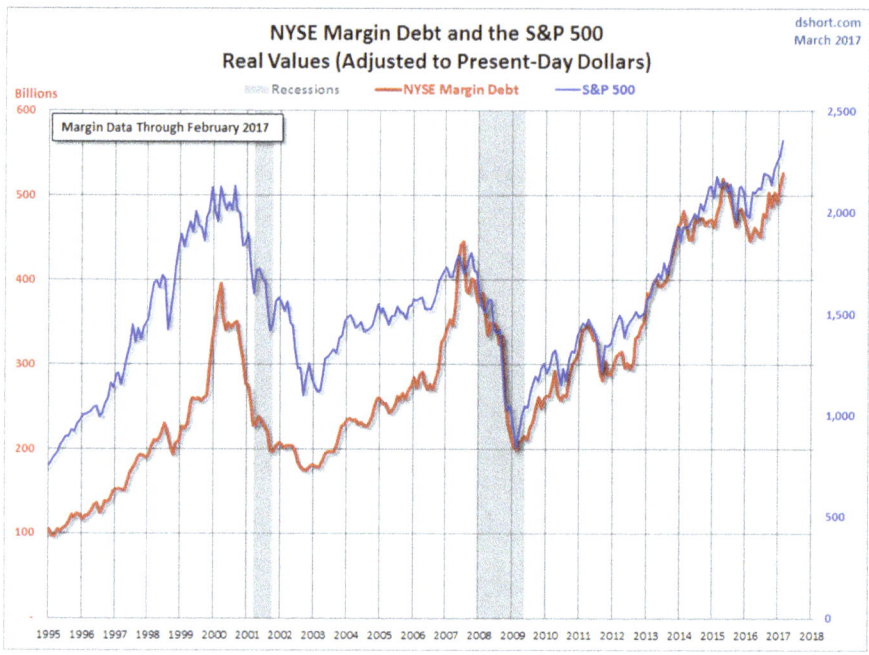

Fig. 1 The use of margin loans to gamble in the stock market

Unequal Voting Rights

Family firms—apart from small shops and farms—are firms that have maintained control by descendants of the founding family by a combination of inheritance and by creating separate classes of shareholders with unequal voting rights. This mechanism allows a small minority of shareholders to control the board of directors. The avowed purpose of this mechanism is to allow founders to maintain control long after they have ceased to make contributions, but another consequence, if not explicit purpose, is to create long-lived dynasties of non-working *rentiers* who benefit from the continuing operations of the firm without contributing anything positive to society. The inequality of voting rights shifts the burden of losses—if the firm is mismanaged—onto all the shareholders, even though only those with special voting rights made the decisions. Tails I win, heads you lose.

Investment Trusts and Margin Loans

An even more effective device for concentrating control over assets in a few hands is called (ironically) "the trust." The first well-known example was the Standard Oil Trust. Trusts, usually created by investment banks, were able to sell their own shares to the public, leaving all decision-making authority in the hands of their sponsoring investment banks. The sponsoring banks were paid a management fee as well as receiving income from the purchase and sale of trust shares.

The business model of the investment banks in the 1920s was to issue shares of existing companies or, leveraged "investment trusts," while the commercial banks promoted margin loans[1] to brokers. The number of such investment trusts in the USA before 1921 has been estimated as about 40, of which US Steel (the mega-merger organized by JP Morgan) was the biggest. But the number grew much faster than the prices of stocks. There were 160 such trusts at the beginning of 1927 and 300 or so a year later. In the year 1928 alone, 186 additional new trusts emerged and in 1929 another 265 were created. The value of shares held in such trusts during 1927 was $400 million. In 1929, the NYSE finally allowed the trusts to be listed. The value of shares in investment trusts sold to the public reached $3 billion in that year, which was at least a third of all capital funds raised during the year. By the time of the crash in October, the trusts had total assets of more than $8 billion (Galbraith 1954, pp. 45–55).

One extreme example was the so-called American Founders Group, organized in 1921 for a total investment of only $500. It grew, mainly by virtue of creating new trusts within trusts, acquiring companies by using stock in a rising market, and maximizing leverage. It finally encompassed a group of 13 companies with total market value by October 1929 of $686,165,000, of which $320,000,000 was invested in other companies in the group. When that house of cards collapsed, it was later found to have owned virtually nothing at all of real value (ibid., p. 64). Money was making money from money (and then losing it) with no underlying substance.

Granted, there are now restrictions that would prevent a repeat of the "American Founders Group" example (at least I hope so). But the practice of using shares as collateral for buying shares in other companies continues, and it is a way of creating "leverage" that amounts to gambling with other people's money. Even though a strategy wins most of the time—say nine times out of ten—there is no justification for "spreading the risk" if the gains are not also shared.

---

[1] A client could purchase 100 shares of XYZ Corp for the price of (say) 25 shares, the remainder being financed by the brokerage (using money from the banks) and held as security. If the value of the shares went up, the customer kept the profits on all the shares. However, if prices went down, the broker would make a "margin call" asking for more security. If the client could not find more money, the broker would sell the shares.

"Greenmail"

A famous example is that of a Texan oil-driller named T. Boone Pickens who created an exploration and drilling company called Mesa Petroleum Co. In 1982, he used the assets of Mesa as collateral for loans that he used to buy enough shares to obtain control of Hugoton Production Co., which was 30 times bigger than Mesa. He bought junk bonds from Michael Milken's firm to do this. Then, of course he used the assets of Hugoton to repay his loans. From there, Pickens did it again, and then again, offering to merge or buy other oil companies, including Cities Service (Citgo) and Unocal. Each time the management would dump assets and do other things to raise the share price—in order to keep Pickens and his investors from controlling the board. Each time, Pickens walked away with a huge profit and a more formidable reputation. Then, in 1985 he attacked Gulf Oil Co., one of the original "seven sisters."

By 1985, Pickens owned or controlled (through nominees) a significant fraction (11%) of the shares of Gulf Oil Company, more than the founding Mellon family holdings. And with that, he literally blackmailed Gulf management (the "commentariat" called it "greenmail") to make radical changes it did not want to make, and finally to sell itself to Chevron. The details of the battle do not really matter here. The point is that Boone Pickens personally walked away with $400 million profit—after paying off his junk bonds—when his Gulf shares were finally sold to Chevron. He is now widely admired by his peers for his philanthropy and public service.

What Pickens and the corporate raiders always threatened to do—and did—was to sell off "unproductive" assets like office buildings and land, cut health coverage, cut retirement benefits, and cut R&D. What this does in typical cases is to increase near term earnings per share, increasing the market value of the shares. But Pickens never actually took control of the companies he attacked. What the Gulf coup did do, actually, was to reduce competition and increase consolidation of the oil industry.

Did Pickens do anything to earn that $400 million by increasing the value of the companies whose shares he bought? No, he did nothing of the sort. (Admittedly the executives in situ share the blame.) In fact, he personally destroyed Gulf as a viable operating company, and as a major employer and supporter of the city and suburbs of Pittsburgh, leaving nothing but a brand. His threat to remodel Gulf in various ways, e.g. by creating a royalty trust, did absolutely nothing to make it more profitable or more productive.

Worse, Pickens also inspired dozens of other corporate raiders (KKR, Carl Icahn, Nelson Pelz, Bill Ackman, Daniel Loeb, Phillip Goldstein, Ivan Boesky, and others) whose combined activities have converted a large part of the US economy from an efficient machine for production and job creation, to a financial casino producing ever larger profits, but at the expense of workers, consumers, communities, and the whole middle class. In effect, the raiders and financial engineers have undone what Henry Ford did a 100 years ago.

Arbitrage, LBOs, and Junk Bonds

The big money game on Wall Street in the 1980s was so-called Leveraged buyouts (LBOs). It was a game inspired by Boone Pickens et al. A corporate "raider" could obtain a "highly confident letter" from Michael Milken's firm, Drexel, Burnham, Lambert, promising financing for a specific project. That was often enough to close the deal. It was how Pickens financed his unsuccessful run at Gulf and Unocal, Icahn financed his attempt to take over Phillips 66, Ted Turner's acquisition of MGM/United Artists and KKR's successful takeover of RJR Nabisco.

"Junk bonds" were bought by savvy investors, usually to finance mergers and acquisition deals. The bonds offered high interest rates to investors because they had no underlying assets except the reputations of the seller. Raiders used the bonds as collateral for bank loans to buy the stock of a target corporation. When the buyout was done, the new owners paid off the junk bonds with assets of the target firm. Michael Milken was the principal broker and seller of the junk bonds, and he also grew enormously rich doing it (until 1988).

One more example (discussed in an earlier chapter) is a firm called Long-Term Capital Management (LTCM), which made use of another device, called *arbitrage*, for making a lot of money from little or no investment, using a lot of leverage. The trick was to find financial assets (usually bonds) selling at slightly different prices but supposedly having exactly the same value at maturity. When there are two prices, and one price is higher than the other, a very nearly fool-proof scheme is to bet that the "true" price is in between the higher and the lower one. So the strategy was to buy the cheaper bond "long"—assuming the "that it would eventually rise in price—and to sell the higher priced one "short," assuming that it would eventually decline in price. The trick is to make sure that the longs and the shorts are exactly matched, so the money received by selling the borrowed "shorts" is exactly equal to the cost of the "longs" in the portfolio. Hence, very little net cash (but a lot of faith) is required.

The most spectacular example of a money-making scheme based on arbitrage was Long-Term Capital Management (LTCM), discussed earlier.

Bailouts, But Only for the Banks

The point is that the losses had to be "socialized" among the banks—all of which were later "bailed out" after the 2008 financial crisis. For some reason, Wall Streeters today blame President Obama for the unpopular bailout, even though it took place under George Bush's Presidency and was organized by his appointees. Curious.

Another way of "socializing" losses from bets using other people's money is illustrated by what happened after the "sub-prime" crisis of 2008. The big financial center banks were buying literally millions of home mortgages, of unknown quality[2] from the mortgage brokers and from FNMA. What they were doing in the years before 2007 was to "package" them as mortgage-based securities (legally they were not bonds), to be sold subsequently to pension funds, insurance companies, and other long-term investors. The banks wanted insurance, for themselves (not their clients), since the mortgage-based securities in their possession represented a large amount of equity, hence a component of the bank's reserve.

As it happened, the Insurance giant, American International Group (AIG) had just the product: the Credit Default Swap (CDS), which was a financial derivative like an insurance policy, in that when an insured object (such as a mortgage-based security) could not be sold, the purchaser of a CDS on that security would be repaid. The CDSs were not insurance policies, however, because the purchase of the CDS on an object need not own the object being insured. Also, there was no limit on how many CDSs could be sold on a given object (security) and the CDSs themselves could be traded. During the years before 2008, AIG made a lot of money—15% of its profits, I think—selling those derivatives. It was like free money, they thought, because they never expected any defaults to occur.

So AIG failed to create a reserve for payments to clients, in case those mortgage-based securities did actually default. And that is what happened.

---

[2] The quality of those mortgages was supposedly assured by one of the three ratings agencies, Moody's, S&P, and Fitch. But, as we now know, the ratings agencies were essentially working for the banks, which meant "looking the other way" when it came to evaluating the risks of the securities the banks wanted to sell.

AIG could not pay. On September 16, 2008, AIG got an $85 billion 2-year loan from the Fed (without collateral) to avoid bankruptcy. The executives who allowed this to happen—indeed they were co-conspirators—walked away with their bonuses. Nobody went to jail. It was another case of socialization of private losses. There were quite a few others, mainly banks plus GM.

There were no bailouts for the homeowners who found themselves "under water" as a result of the massive house price decline that followed the subprime crisis.

Shareholder Value Maximization (SVM)

Shareholder value maximization (SVM) is a theory of management. It is also a new doctrine in economics. It is most often attributed to Milton Friedman, who said in *New York Times Magazine* back in 1970: "*There is one and only one social responsibility of business—to use its resources and engage in activities designed to increase its profits*" (Friedman 1970). He also said that corporations are not "persons" (true), even though corporations have the legal status as persons. It follows, in his view, that shareholders necessarily (being rational utility maximizers, by assumption) want to maximize profits. Hence any act of corporate social responsibility is tantamount to "*taxation without representation.*"

There is an ad on CNBC, showing a presumably wealthy investor who says he does not drink coffees with fancy names. He says: "*Every day I drink a cup of coffee. It costs 18 cents to make. I invest the rest. It works.*" This TV investor does not say what he will do with the money he is investing by not consuming fancy coffee or other "crap" (his word). The viewer is supposed to think that making as much money as possible is an end in itself. Most of Wall Street agrees with that sentiment.

This fits the new "principal-agent" theory of economics. It assumes that corporate managers are simply agents of the owners (shareholders), whence SVM seems to follow automatically (Jensen & Meckling 1976). However Lynn Stout (among others) has pointed out that, notwithstanding the claims of many "activist" investors, corporations—being legal "persons"—are required to act in their own best interests (as determined by their Boards of Directors) and are *not* required to act as fiduciary agents for the "owners" (Stout 2012). This important legal difference has been confirmed a number of times in the courts.

Notwithstanding legality, the activists have taken over most of the corporate board rooms, at least of public companies. It is pertinent to note that in

1981 the *Business Roundtable* said *"Corporations have a responsibility, first of all. To make available to the public quality goods and services at fair prices, thereby earning a profit that attracts investment ... provide jobs and build the economy"* (Montier 2014, p. 2).

James Montier went on to compare the performance of two corporate icons, IBM and Johnson & Johnson (J&J), in total returns to shareholders after 1973. Up to 1988, both firms were managed more or less according to the dicta of the 1981 Business Roundtable, and they were roughly equal in terms of performance.[3] The guiding principles of IBM, as recapitulated by Chairman Thomas J. Watson Jr. in 1968, were (1) respect for employees, (2) commitment to customer service, and (3) achieving excellence in all domains of business. Shareholders were not mentioned. Similarly, the mission statement of J&J, written by its founder R.W. Johnson in 1943, said (and still says) the following:

> *We believe our first responsibility is to [all] ... who use our products, ... We are responsible to our employees ... We are responsible to the communities in which we live and to the world community as well ... Our final responsibility is to our stockholders ... When we operate according to these principles, the stockholders should realize a fair return.*

Yet a decade and a half later, the idea of corporate responsibility to customers, workers, or communities was out of the window. The Business Roundtable in 1997 pronounced that *"The principal objective of a business ... is to generate economic returns to its owners ... if the CEO and the directors are not focused on shareholder value, it may be less likely the corporation will realize that value"* (Montier, *op cit*).[4] Behind those words is another assumption: i.e. that share price is the best measure of shareholder value.

Consider IBM again. From 1990 on, IBM was run by a series of SVM advocates starting with Lou Gerstner. And since 1990, it has lagged far behind J&J in relative terms. In 1974, IBM, under the Watson regime, was the fifth most valuable company in the USA, while J&J was not in the top 20. By

---

[3] Confession: I was a summer intern at IBM in 1954, as a programmer. At the time, IBM was getting started in the computer business under T.J. Watson Sr. The company motto then was "Think". In those days, employment in the company was regarded as an honor, and employees were paid well and provided with exceptional benefits.

[4] The title of Montier's paper about SVM is "The World's Dumbest idea". He credits those words to the legendary Jack Welch of GE in an interview with Financial Times in March, 2009.

1994, IBM had climbed up to fourth place, after GM, Ford and Exxon-Mobil, and J&J was still not in the top 20. But by 2014, J&J was number 5 while IBM had fallen back to 13th place. This cannot be explained by IBM being in the wrong industry, considering that Microsoft, born, in the 1980s, as a stepchild of IBM's PC development, had reached fourth place on the list by 2014 (with Google in third and Apple in first [Barry 2014]).

What went wrong at IBM? The short answer is PCs. Dedication to "mainframe" computers and failure to respond adequately to the challenge posed by personal and desktop computers made by DEC, Compaq, Apple, and others was the first cause of its fall from grace. IBM's decision to abandon manufacturing altogether may have been the second. (Why did IBM fail to acquire Microsoft when that would have been so easy?)

But since the 1990s, blind dedication to SVM (which continues) has led to unending emphasis on cost cutting (by job cutting), lack of product innovation, and use of cash to finance corporate stock buybacks. Between 2005 and 2014, IBM delivered $32 billion in dividends to shareholders and spent $125 billion buying its own shares (to prop up the share price), while investing only $111 billion in capital and R&D, together. IBM today is a sad shadow of what it was in the 1950s, '60s, and '70s. Whether a long fixation on artificial intelligence can save the day remains to be seen.

This sad story is not unique to IBM. The recent history of Hewlett-Packard (HP), the first of the great success stories in Silicon Valley, is equally depressing. When Carly Fiorina took over in 1999, she started a share buyback program. During her term (until 2005), HP bought back $14 billion in its own stock, while earning only $12 billion in profits. (It should have invested in Apple.) Under the next CEO, Mark Hurd, HP paid $43 billion for its own stock, while earning only $36 billion in profits over 5 years. The pattern continued under Leo Apotheker ($10 billion in stock repurchase) and Meg Whitman, his successor in charge, who broke up the firm by spinning off printers from software. By the time HP's current "turnaround" is complete, it will have cost 80,000 jobs (Brettell, Gaffen, and Rohde 2015)

"Buybacks" were illegal throughout most of the twentieth century because they were considered a form of stock market manipulation. But in 1982, the new Chairman of the Securities and Exchange Commission (a Reagan appointee) introduced a new rule (known as 10-b 18) that created a legal process for buybacks without Congressional approval. This opened the floodgates for companies to start repurchasing their stock *en masse*, usually to increase the share price—benefitting shareholders and option-holders—by reducing the number of shares in circulation. Today, when CEOs do not need to use the firm's current profits to build more factories, or develop new products, they invoke the "shareholder value maximization" (SVM) mantra and engage in share buybacks.

The goal of SVMs, to the exclusion of all other goals, is now widely adopted by corporate boards and taught in business schools.[5] It underlies the field of "private equity," i.e. the systematic privatization of public companies originally created with different goals—e.g. to create products and serve customers—into single-purpose profit maximizers.

Nearly 60% of non-financial public companies in the USA have bought their own shares since 2010. In the last reporting year (2015), share repurchases were $520 billion, along with $320 billion in dividends, adding up to $885 billion, as compared to net income of $847 billion (op cit.) Annual total real returns of US public companies (as percent) from 1940 to 1990 were about 7% p.a. However, real returns since 1990, allowing for the share price rises due to supply reduction due to share buybacks, is barely 5% p.a. (Montier 2014, Exhibit 3). This is why pension funds and insurance companies are now in deep trouble since they mostly predicated their pension and insurance offerings on a continuation of that 7% history.

In summary, what the SVM movement has done since 1982 is to shift a large fraction of corporate profits into asset purchases (the S&P) and away from long-term investment in new products and services. The implications for future economic growth are not favorable. At the same time, the continuous focus on SVM since 1982 has undoubtedly played a major role in driving inequality by pushing wages down and corporate profits up to record levels.

The context of Friedman's article is forgotten, but relevant. The headline of the article in NYT Magazine where Friedman said "*The social responsibility of business is to increase its profits*" was followed by a photo of GM Chairman James Roche at a stockholder's meeting. He was replying to members of "Campaign GM," a national group that had demanded that GM should name three new directors to represent the "public interest." Not surprisingly, GM management hated the idea and campaigned fiercely against it. Friedman's article was one of his weapons. The GM shareholders, mostly institutions, loyally rejected the proposal. That event was probably the beginning of the decline of GM. GM management won the battle over (not) including representatives of the public interest on their board. But they simultaneously put themselves on the wrong side of several controversies, not only safety but also air pollution, fuel economy, and ease of parking. The US carmakers wanted to make big "fuel guzzlers" (they still do) because they are more profitable than small cars.

---

[5] In an email answer to a recent question from Reuters, Itzhak Ben-David, Professor of Finance at Fisher College of Business, Ohio State University, said "Serving customers, creating innovative new products, employing workers, taking care of the environment … are NOT the objective of firms. These are components in the process that have the goal of maximizing shareholder value" (Brettell, Gaffen, and Rohde 2015).

However, contrary to Milton Friedman's assertion, corporations are "legal persons." Corporations legally own themselves. They are not actually owned by—or responsible to—shareholders except indirectly insofar as shareholders vote for directors, who appoint management. This has been confirmed a number of times in the courts, including the US Supreme Court (at least with respect to political contributions). The courts have ruled that Boards of Directors are required to act in the "best interests" of their client (the firm). *But directors and managers do not have any direct fiduciary responsibility to the shareholders* (Stout 2012). The brokers, hedge funds, and other institutional investors, backed by academics, paid no attention to the court's rulings.

Apart from Friedman's misunderstanding about the legal rights accruing to share-ownership, Friedman's narrow view of corporate responsibility led to a mini-revolution in economics, known as the "principal-agent" theory. This was elaborated in two influential academic papers on the theory of the firm, by Michael Jensen and William Meckling (Jensen & Meckling 1976; Jensen 1884). The two authors said they wanted to devise a simple objective function for managers—as opposed to vague "mission statements" such as "public interest," or multiple objectives such as "triple bottom line" that cannot be mathematically maximized.

To simplify the problem, Jensen and Meckling assumed—from the beginning—that everything of importance in a business can and should be monetized. Consistent with Milton Friedman's ideas, they saw no need for businessmen to maximize employee loyalty, consumer satisfaction, community welfare, or environmental protection unless they affected the "bottom line." They assumed that only shareholder's ("owners") need be satisfied. There is no need (they said) to satisfy customers, employees, neighbors, or the national interest. Having made those (quite radical) assumptions, they argued that an ideal "objective function" for a manager must be *to maximize the return on capital invested in terms of free cash-flow relative to a risk-adjusted "hurdle rate," i.e. based on what investors "expected."*

They proposed this rule despite its essential ambiguity: It is mathematically impossible to simultaneously maximize any objective function at two different future times, and it makes no sense to maximize over a period of time. Yet generations of MBA students, notably at Harvard Business School (where Jensen taught), were persuaded that this was the golden rule for managers. However, this theoretical "objective function" is impossible to calculate in practice for day-to-day decision-making. Real executives cannot use it. What they can (and do) do is to ask themselves whether a given action is likely to raise or lower the market price of the stock. Hence, it is the *current price of the shares*, rather than the overall value of the business, that has become the test of managerial performance. It is a very poor test, for a variety of reasons, including the fact that—because of it—managers are now obsessed with short-term performance, and companies that are not run by the owners (I do not mean the shareholders) increasingly neglect long-term considerations.

Share Buybacks Vs. R&D

Inspired by Schumpeter, the work of Mensch et al., among others, argued that weak consumer demand, governed by the business cycle, enables increased levels of industrial R&D resulting in more product innovation (Mensch 1979 #8003, 1988 #8173). There is some evidence for this behavior in the past. But in recent decades, it appears that the classical R&D investment mechanism in corporations has been short-circuited. When revenues and profits declined because of economic slowdowns, it seems that profits are less and less retained and diverted to R&D, or other long-term objectives as Mensch suggested. Instead they are now mostly returned to shareholders, via buybacks. Most economists seem to think that this is perfectly rational behavior, allowing the "free (financial) market" to re-allocate capital into the most profitable opportunities. Maybe this is true if financial sector profits are counted, but for "brick and mortar" sectors that produce goods and non-financial services, the evidence is against that view.

When companies first make an initial public offering (IPO), stock is sold to the broad public on one of the main stock markets. In principle, the sale of the stock can be used to raise money to finance further growth. More typically, the secondary reason for the IPO is to give the early investors in the company—venture capitalists, early employees, and founders—the opportunity to "cash out," i.e. be repaid, in part, for their risky investments. A stock buyback is basically a secondary offering in reverse—instead of selling new shares of stock to the public to put more cash on the corporate balance sheet, a cash-rich company expends some of its own funds on *buying* shares of stock from the public.

What does this mean for the economy as a whole? From 2008 to 2017, 466 of the S&P 500 companies distributed $4 trillion to shareholders as buybacks, along with $3.1 trillion as dividends (Lazonick 2014). The 10 biggest share buybacks during 2003–2012 added up to $859 million.[6] The average CEO pay for those firms was $168 million, while the next four highest paid

---

[6] The "top 10" in order were Exxon-Mobil, Microsoft, IBM, Cisco Systems, Procter & Gamble, Hewlett-Packard. Walmart, Intel, Pfizer, and GE.

executives got $77 million each. The CEO pay was 58% based on stock performance (options and awards), while the other top executive pay was 56% based on stock prices. Incredibly, *all but three of those companies spent more than their net income on buybacks.* Hewlett-Packard (177%) was the highest on that list, followed by Pfizer (146%), Microsoft (125%), Cisco Systems (121%), Procter and Gamble (116%), IBM (111%), and Intel (109%). The lowest of the ten was Walmart (73%).

The $1.5 trillion Republican Tax Cuts and Jobs Act of 2017 slashed the corporate tax rate to 21% from 35%, reduced the rate on corporate income brought back to the United States from abroad to between 8% and 15.5% instead of 35%, and exempted American companies' foreign income from US tax. Since the tax bill, companies have been doing a lot of buybacks. Share buybacks in 2018 averaged $4.8 billion per day, double the pace from the same period in 2017.

Goldman Sachs estimated that after the 2017 tax cuts S&P 500 firms would return $1.2 trillion to shareholders via buybacks and dividends in 2018, increasing share buybacks by 23% to $650 billion. A Bloomberg analysis found that about 60% of tax cut gains would go to shareholders, compared to 15% to employees. A Morgan Stanley survey found that 43% of tax cut savings would go to stock buybacks and dividends, while 13% will go to pay raises, bonuses, and employee benefits. Just Capital's analysis of 121 Russell 1000 companies found that 57% of tax savings would go to shareholders, compared to 20% directed to job creation and capital investment and 6% to workers. By one tally, the tax cut scoreboard stands as follows: Workers $6 billion; Share buybacks $171 billion.

There is another kind of evidence against buybacks. In the next few lines, you will see the results of unpublished research on this topic at INSEAD, conducted by me, jointly with my colleague, Michael Olenick (Ayres & Olenick 2017). We compiled a list of all corporate buybacks between January 1, 2000, and December 31, 2016, by US firms trading on the three major exchanges, with a market cap of $100 million or more at the latter date. This yielded 5,448 share-purchase transactions by 1,015 US companies. For each business, we computed its total expenditure on buybacks, during that period, adjusted for inflation. We also compiled the data on current market value for each company, as compared to its market value 5 years earlier. We then computed, for each company, the ratio of inflation-adjusted buybacks to current market cap as a percentage. That is, if a business had a market cap of $100 million and repurchased $50 million of shares of that business would have a ratio of 50%. We then rank-ordered them and plotted these percentages against the 5-year growth percentage.

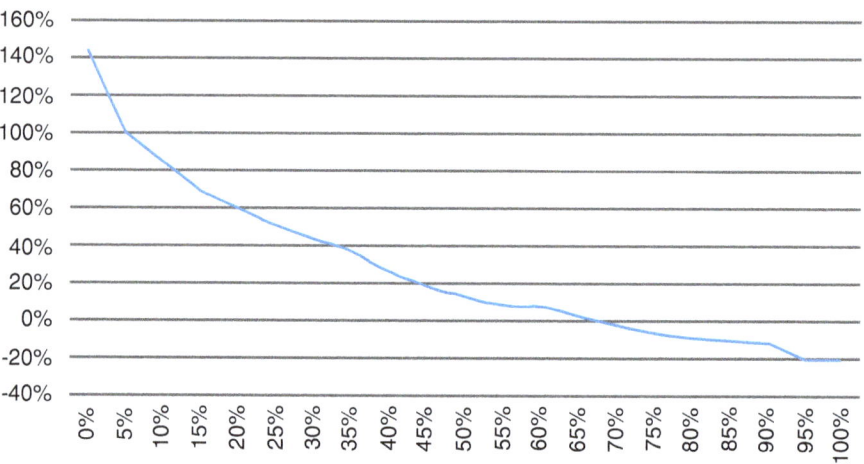

Fig. 2 How stock buybacks affect company growth

The results of our research are shown graphically in Fig. 2. The vertical axis is growth; the horizontal axis is the ratio of buybacks to market value at the end of 2016. The trend is clear. The more the firm engaged in buybacks, the less it grew. The more a company spent on buybacks, the less it grew, in market value, over a 5-year time-scale. Companies that have spent a large fraction of their current market cap on buybacks are virtually guaranteed to decline in the coming years. The 535 firms that repurchased less than 5% of their market value as of December 31, 2016, saw their market value increase an average of 247.8% over the prior 5 years. The 64 firms that repurchased 100% or more of their final market capitalization experienced a 21.7% decline in value over the same timeframe. Note that the curve must continue further and lower beyond the 100% point, although we did not bother to plot it.

As mentioned, market cap growth is based on the 5 years 2012–2016. Five years was used as a benchmark because recent data is more reliable, and many firms actually disappeared during the "crash years" making interpretation of the results much more difficult. The strength of this correlation relationship can be tested statistically, and it is obviously (by the eyeball test) extremely strong. For statistics mavens, the chi value is 0.02544, which means the chances of non-correlation are negligible. In other words, there is a strong causal relationship between buybacks and growth rates. Indeed, the graph suggests that companies that engage in excessive buybacks—beyond 50% or so of market cap—will almost certainly lag behind the S&P in growth. Firms where buybacks exceed 65% of market cap are likely to experience zero net growth over 5 years, and beyond that point absolute declines can be expected. Large wealth management funds and pension funds should be using this information in making long-term investment decisions.

What does all this mean for the economy? For one thing, it means that firms that have "invested" in buybacks (to support the price of the stock and to keep the senior executives happy) *have actually wasted money that should probably have been invested in the business, especially in R&D.* It is also clear that buybacks make inequality worse, thus reducing opportunity. Barely half of Americans own stocks at all, and what is more, the richest 10% of Americans own 80% of all traded shares. The bottom 80% of earners own just 8% of shares.

William Lazonick recently told a CNN Money audience: "*Buybacks have been a prime mode of both concentrating income among the richest households and eroding middle-class employment opportunities.*" Congress is taking notice: the Reward Work Act introduced by Senator Tammy Baldwin in March 2018 would ban stock buybacks done as open market repurchases. (It would still allow buybacks done through tender offers, which are used for different purposes.) The chances of passage seem minimal as long as the Republicans hold the Senate.

Leverage: How the Rich Keep Getting Richer

All of the mechanisms cited in the previous chapter are sources of increasing inequality. But I think that the common factor is a "trick" that is a common feature of western capitalism, although it is not fundamental to the idea of investing savings or making a profit. This trick is called *leverage* or more accurately, *gambling (making bets) with other people's money*. Thanks to the limited liability laws now on the books, the gamblers who win, making investments by borrowing and using other people's capital (usually through corporations), get to keep most of their winnings, giving very little (if any) back to the people whose money they are renting. Even the winnings—capital gains—are taxed at a very low rate.

On the other hand, when those well-connected gamblers lose their bets, the losses are "socialized" by being suffered first by the savers whose money was malinvested, and then by the whole society ("the taxpayers"). This seems inconsistent with Rousseau's conception of a *Social Contract*. The view that the business of business is to make profits for the shareholders, and nothing else, was stated succinctly by Milton Friedman: "*the social responsibility of business is to increase its profits*" (*New York Times* Magazine, September 13, 1970). This widely quoted statement has been partly, if not totally, responsible for significant changes in the conduct of the capitalist "game", starting with Ronald Reagan's election in 1982.

Reverting to the Reagan years, the first major change was the "leveraged buyout" (LBO) craze, pioneered in 1982 by William Simon, formerly of Salomon Brothers, and later Secretary of the Treasury. Simon and his partners invested $1 million of their own money (as Wesray Capital), plus money from "junk bonds," backed by the assets of a target company. Their idea was to buy

Gibson Greeting Cards from the Radio Corporation of America (RCA), (which had just been bought by General Electric, in order to acquire NBC). RCA was not interested in Greeting cards (why did they buy it in the first place?).

So RCA sold Gibson to Wesray Capital for $58 million in cash (using other people's money) and Wesray took over $22 million in liabilities. The money came from junk bonds collateralized by Gibson's assets, not by Wesray's assets. That was possible only because of William Simon's personal reputation. (You or I, dear reader, could not borrow a lot of money based on a promise of repayment from the profits of a future project.) But Gibson Greeting Cards was bought by Wesray, and after some financial engineering (cost cutting), it was resold to the public via an Initial Public Offering (IPO), for $330 million. William Simon made a personal profit of $66 million from a personal investment of $333,000 (Ayres 2014). How's that for leverage?

Needless to say, this example of the power of leverage, for buyouts, inspired many imitators. It spread during the Reagan era when "corporate raiders"—e.g. Marvin Davis, Asher Edelman, James Goldsmith, Carl Icahn, Irwin Jacobs, Nelson Peltz, Ronald Perelman, H. Boone Pickens, Kohlberg, Kravis and Roberts (KKR), Saul Steinberg, and Ted Turner—using borrowed money (mostly from junk bonds), took over companies via leveraged buyouts (LBOs) and "financially re-engineered" them. Once Ted Turner quipped, "*It's not how much you earn but how much you owe.*" At the time, he owed $2 billion, which he borrowed to buy MGM. (He got it back by selling off MGMs archive of 3,800 old movies.)

The LBOs in the 1980s were financed by high yield "junk bonds." The junk bond business was pioneered by Michael Milken at Drexel, Burnham & Lambert. In 1986, Drexel made a profit of $545 million and became the darling of Wall Street. In 1987, Milken paid himself $450 million. Later, Milken went to jail for a different sin—insider trading—and Drexel itself failed in 1989, partly due to Milken's actions. But the leveraged buyout activity has continued. It is now a recognized branch of the financial sector, rebranded today as "private equity."

The main purpose of "private equity" groups is to undertake financial engineering of target companies, and to avoid the regulatory oversight that public companies require. The financial engineering procedure advocated by legendary promotors, like Bruce Wasserstein and Joseph Perella (working for First Boston Corp.), was to issue a "wake up" call for "sleepy companies" with "lazy managers." The new owners would sell off real-estate or other "unproductive" assets (such as office buildings, land, new product development, or long-term R&D) lay off older workers approaching pension age, replace "defined benefits" pension plans with "defined contribution" plans, and squeeze the remaining employees in other ways (cutting medical benefits, for instance).

The "new" more profitable company was then relabeled and sold to the public—like Gibson Greeting Cards—at a much higher price by means of an IPO. Or, after a leveraged buyout (LBO), the junk bonds were paid off by the company itself, using more conventional bank loans. This greatly increased aggregate corporate debt during the 1980s. The increased use of corporate debt was justified to the public by the fact that corporate profits used for dividends are "double taxed," while interest paid on corporate debt is tax deductible. The game became known, informally, as "buy, strip, and flip."

Companies bought and flipped in this way became more efficient, in a narrow corporate sense, but they did not become more productive in any sense. The raiders like Simon and Pickens got richer and the workers lost their jobs and savings. The game is no longer played quite as it was in the 1980s because Boards of Directors of vulnerable companies have adopted a number of gimmicks like "poison pills" and "golden parachutes" to protect themselves from hostile takeovers. Takeovers today are usually consensual. Consent to merge is bought by making deals with large shareholders or share buybacks (discussed in the last chapter). But the ideology of maximization of shareholder value maximization (SVM)—at the expense of workers, customers, and other stakeholders—has continued to spread, especially as it is now taught in business schools. The enormous pay packages for chief executives and their immediate subordinates are based heavily on stock-market-based incentives, such as options that kick in when the share price moves above a specified level. It is no accident that stock-market valuations have risen much faster than the economy as a whole since 1980, while wages in real terms have lagged or dropped. Shareholder value maximization and rising stock prices do nothing for the employees, the customers, or the taxpayers.

Fractional reserve banking originated (under other names) in the eleventh century, by the Templars, as explained in the chapter on Credit and Banking. Reserves back then were gold. Today gold is no longer the base for currency, and it is possible for banks to lend considerably more than they have in deposits or keep in reserve, based simply on the "law of averages." Big banks in 1998 typically kept reserves of around 6% of loan value, corresponding to a leverage multiplier of 15 or 16. By 2007, the leverage of the big banks and brokers in New York and London had risen to the range of 25–33. Now, according to the so-called Basel rules,[1]—responding to the immense losses of 2008–2009—the biggest banks are required to keep significantly larger liquid monetary

---

[1] Basel rules are set by a consensus of national banks, under the auspices of the Bank of International Settlements (BIS) located in Basel.

reserves, far exceeding the global gold supply (and yet they complain). There is a lot of argument about what counts as a liquid reserve and what does not. However, I shall skip that question, except to note that the financial crash of 2008 was triggered by a change in the "liquidity" status of mortgage-based bonds or Collateralized Debt Obligations (CDOs).

On average, what the banks do is admittedly safe, most of the time. But the law of averages is a statement about probabilities. If an event has a probability of 90%, there is still a probability of 10% that it will not happen. Or, if it has a probability of 90% of not happening, there is a 10% probability that it will happen. It follows that, sooner or later, some bank will suffer a "run" when its depositors all want their money back at the same time. When that event happens, the bank will fail. This has happened many times in the past when a "bubble," such as the South Sea Bubble, collapsed.

It happened to the Continental Illinois National Bank (CINB)—then the seventh largest bank in the USA—in 1984. The trigger was the failure (in 1982) of Penn Square, a much smaller bank in Oklahoma that specialized in lending to oil and gas prospectors. Those prospectors were drilling too many "dry holes." Penn Square owed a billion dollars to CINB, which it could not pay. Depositors in CINB got nervous and began to move to their assets to other banks. Eventually the Federal Deposit Insurance Corp. (FDIC) had to step in. FDIC paid out $4.5 billion, and CINB itself was swallowed up by Bank of America. All of this happened because of surplus of dry holes in Oklahoma.

One of the reasons the ultra-rich get richer is that they can (and often do) create "bubbles" for profit. Such bubbles entail large wealth transfers as the asset value goes up and later falls back down. The winners are the early investors who get out early. The losers are the latecomers who borrow to get in and do not get out soon enough, or saddest of all, the bystanders whose assets have (unknown to them) been used as collateral for defaulted loans to others.

Excuse a brief digression: In card games, games of dice, like poker, there is a loss for every gain. The sum total of losses and gains, in a poker game between friends, is always equal to zero: hence it is a "zero-sum" game. At horse races, dog races, automobile races, and so on, there is a lot of betting. But, because of the costs of the gambling establishment—taking bets, computing odds, and paying off the winners—the total of losses to the bettors virtually always exceeds the total of winnings. The betting shop takes its cut from the winnings, of course. This makes it a "negative-sum" (lose–lose) game for the players (bettors). There is an aggregate transfer of wealth from the bettors to the betting shops, no matter which horses or cars win the races.

OK, the gambling establishment at a race track provides a service to the customers. The service is entertainment and excitement, and it presumably justifies its modest cost (on average) to the punters. But the fact remains that the total of punters losses exceeds the total of other punters winnings. The same thing is true when stocks and shares are traded on the stock market. The total of the winnings by investors in the stock market on a given day is always less than the sum of trading losses by investors, unless "new money" comes into the market. The difference consists of fees by the stockbrokers, fund managers, and investment banks that trade stocks and shares for clients.

Even if the sellers and buyers are essentially the same people, there is a net transfer of wealth from individual or corporate investors (buyers and sellers) to the intermediates: namely the stockbrokers, fund managers, traders, and the banks that employ them. The stock market is a negative-sum (lose–lose) game for the investors as a group. Not many small investors consistently make money. Those who do make money (like Warren Buffet) did so by being good macro-trend forecasters or by exploiting information asymmetries. By that, I mean that the winners have access to earlier—and more accurate—data than you and I do, or they are better at interpreting the data we all get.

The situation is fundamentally different for investors who put their money directly into "hard" assets, such as houses, rental properties, or into new ventures. (This may be why the venture capitalists cluster around Silicon Valley or Route 129 around Boston, and not around Wall Street.) Of course, there is still a need for brokers and credit, but there is a subtle but important difference between investing in a share of a real new venture (or a new house) and simply buying a share in an existing one. The winners from investments in new ventures may be relatively few, *but their winnings are not necessarily somebody else's losses*. Moreover, the winnings can be much larger than the losses to unsuccessful ventures, i.e. the game can be *positive sum* (win–win). It is important to understand this distinction because overall economic growth is mostly driven by the relatively small number of win–win bets, not the large number of win–lose or lose–lose bets.

Having said this, it is also important to realize that the great majority of trades on Wall Street are bets in zero-sum games, where only the intermediates (bankers and brokers) are guaranteed winners. The "investment bankers" supposedly allocate capital (other people's savings) into the most profitable sectors. This is a myth. What they really do is to manage merger and acquisition (M&A) deals, in exchange for large fees. If the deal is "successful," the combined enterprise is more profitable than its original components, and the bankers and "private equity" investors win. When this happens, it is mostly the result of eliminating duplicate or unnecessary functions, resulting in job losses. Yet many, if not most, of such deals are not successful, ending in a spinoff or a "fire sale." The investment bankers do not care, because they do not risk their own money and even undoing a former deal is also a source of large fees. The role of the investment bankers and brokers is essentially parasitic, except where they bring out a successful Initial Public Offering (IPO) of a new venture.

For example, the role of the banker-brokers in arranging the sale of BP Group to Shell, a few days before I wrote these words, added nothing to total global wealth. It was a zero-sum game. Gains (to some) matched losses (to others). The outcome gave the bankers big fees, gave Shell more gas reserves, and eliminated one of its major competitors in one market. The employees and shareholders of BP will be major long-run losers. The same can be said of the role of the banker-brokers in restructuring General Electric Corp. (GEC) by selling off its real-estate and other financial assets. The fact that a large portion of the cash generated by that sale will be re-invested in the shares of GEC—and not in new products or projects—makes the point. It is a triumph for the maximizers of short-term shareholder value, and it will temporarily increase the wealth of the owners of those shares, but cashing out its past profits in this way will decrease the future value of the company: it is a lose–lose game for the long-term investors and for the economy.

The growth of the parasitic financial sector *vis-a-vis* the rest of the US economy in recent years makes the point in yet another way. It is not surprising that the financial titans work hard to find new ways of privatizing their gambling gains while "socializing" their losses. The problem is that the regulators are always "behind the curve." The financial collapse of 2008 was truly a "near death experience" for the global financial system. It was caused by two things. One was the invention of a financial gimmick called "securitization." The other was a real-estate bubble.

The securitization gimmick (which I will not try to explain here) enabled banks to construct bonds with fixed maturity and fixed interest rates from large numbers of heterogeneous home mortgages. Back in 2003–2007, the banks persuaded the three credit ratings agencies (Moody's, S&P, and Fitch) that these bonds deserved AAA ratings. That made them suitable for sale to pension funds, college endowments, insurance companies, and other long-term investors catering to "widows and orphans." The banks were able to make large profits on these sales, while simultaneously transferring all but a small fraction of the risks to their clients without keeping any of their own "skin in the game."

The crash of 2008 was actually triggered by a rise in mortgage default rates in the spring of 2008. This triggered a reduction in the credit ratings of many AAA-rated mortgage-based securities, making them suddenly—overnight—unusable as bank reserves. The other cause of the financial collapse of 2008 was the real-estate bubble itself. Both the Clinton and (George) Bush administrations wanted to increase homeownership as a social goal. They pressured the mortgage industry to make housing more available to low-income people.

The obvious way of doing this was to loosen the standards for evaluating home-buyer credit applications. This could be, and was, done administratively, without Congressional knowledge or approval. No new laws were needed. Then when the mortgage-based bonds proved popular to long-term investors, the private mortgage sellers (The Money Store, HFC, Famco, Option One, New Century, First Plus and Country-wide) began to offer 100%, adjustable rate mortgages (ARMs) to anybody who could sign their name, no job necessary, no money down. Again, the bankers were not taking any risks with their own money

For a while, this tactic paid off in a big way. It increased housing demand and drove house prices up still further. This put more money in the pockets of the home sellers, the housing construction industry, and the mortgage sellers (who sold, in turn, to banks that "securitized" and packaged them as bonds). Some home loan mortgages were sold to the government sponsored (but privatized) mortgage finance institutions Federal National Mortgage Association (known as FANNIE MAE) or its cousin Federal Home Loan Mortgage Corporation (FREDDIE MAC). Those organizations were created by Congress back in the Great Depression and early post-war period to finance housing for the ordinary man-in-the-street.

Fast-forward to 2006–2007, buyers with no equity at risk could (and some did) buy several houses (to rent), and refinanced them when the ARMs were due to kick in. Some of those people later just walked away—defaulted—leaving the mortgage-based bond-holders—pension funds, insurance companies, etc. "holding the bag," so to speak. The real losers were the pensioners and others depending on those funds.

And then, in September 2008, the bubble burst. Congress created the Troubled Asset Relief Program (TARP), in a hurry. Except for Lehman Brothers, the too-big-to-fail (TBTF) banks, and two other mega corporations were bailed out. American International Group (AIG), the world's largest insurance company, which made a lot of money selling credit default swaps (CDSs) got the biggest bailout of all. Home equity losers got no bailout.

Overall, the loss of home equity in the USA was $8.2 trillion or $5.5 trillion, allowing for the $2.4 trillion that some homeowners took out in cash and spent during the price rise. At the low point, homeowners still had some equity (about $3.8 trillion). But many individual homeowners, who had accumulated home equity in a rising market, suddenly found themselves "under water," owing the banks more than the value of the house. (My younger brother was one of them.) Altogether, 22% of US homeowners found themselves "under water" and people in that group lost all—or even more than all—of their life savings, in the proverbial "blink of an eye."

There were other losers, too. Stock-market losses were also very large. From the peak in 2007 to the trough in 2009, the loss of paper value was about $8 trillion. Most of that loss had been made up by 2017, thanks in part to cheap money and quantitative easing "QE" and, in part, to increasing corporate profits due to the spread of short-term "shareholder value maximization" ideology, and leading to unrestrained mergers and consolidations, as Joseph Stiglitz has pointed out (Stiglitz 2015). Not so for the middle-class homeowners, most of whom were not shareholders. There were a few big winners, like Goldman-Sachs and a few hedge funds, who managed to bet against their own clients. Overall, the middle-class home equity owners (savers) got none of the short-term gains when house prices went up, but took most of the loss when the bubble burst. This one event left the middle class in the USA with significantly less of the national wealth pie. It left the rich even richer, in relative terms. The losses to the middle class—comparable to what happened in Germany during the hyper-inflation of 1922–1923—may have contributed to the surprising election of Donald Trump as President of the United States, in 2016.

One of the myths of modern finance, as taught in business schools and as practiced in corporate boardrooms, is the notion attached to SVM theory that CEOs and Boards have a *fiduciary* (i.e. moral) obligation to maximize returns for shareholders (meaning stockholders), regardless of any other interests. Share-owners interests currently trump the interests of other "stakeholders": worker-employees, customers, third parties affected (taxpayers, the public, the environment). Meanwhile, the owners of debt always seem to trump the interests of equity owners. There is now a hierarchy of superior vs inferior (subordinate) bondholder rights. When a firm is bankrupt, taxes and wages are paid first. The senior debt holders get paid next, then those further down the hierarchy. Equity-holders get paid last (if at all). Debt repayment is still regarded by conservatives as a moral obligation. Consequently, property rights are superior to human rights, which are much less well-defined and less enforceable.

The end of legal slavery meant that people in western countries now "own" themselves. But slavery of the body has been replaced by something else. Companies cannot "own" their employees as such. But by means of legal contracts, they can own their employees time and/or work product, including their ideas and intellectual property. Governments still absolutely own the services of soldiers, by a very one-sided contract ("the King's shilling"). Movie companies used to "own" their employees' faces and voices (remember "Singin in the Rain"?), and professional sports teams still "own" their players, for periods of time, by contract.

Is Capitalism Per Se Responsible for Lifting Billions of People Out of Poverty? Or Is It Responsible for Increasing Inequality?

There is a growing public debate on the state of US corporate capitalism. Defenders of the *status quo ante* dominate the media outlets such as CNBC, Bloomberg, and Fox News. They are mostly allied with the Republican Party. With few exceptions, they tend to regard "capitalism" as a belief in the virtues of private ownership of the means of production, with "free markets," meaning little or no regulation by government. By "free markets," they mean markets with minimal government interference, mainly to defend the country, enforce contracts, protect property rights, and maintain "law and order."

Those who call themselves "capitalists" in this debate tend to distrust "big government" and characterize their opponent's beliefs as "socialism." By that epithet, they do not mean Marxism as such (which few understand). They say that "socialism" is "big government," including welfare programs for the poor and active government oversight and interference in markets. What unites them above all is opposition to "redistribution" from the rich to the poor, and dislike of income or wealth taxes. (They are happy with sales taxes and "flat" taxes.)

Even in the so-called liberal wing of the Democratic Party, there are very few if any—actual socialists in the classical sense, i.e. who believe in government ownership of the means of production, still less the communist mantra "from each according to his ability, to each according to his need." In fact, I know of none who do. That idea, espoused by some idealists around the time of the French Revolution, has rarely been tried in practice, and has never succeeded over a long period of time. So the argument today is really about the shape and size of the social "safety net" in particular. I have a specific proposal on that, explained in the last chapter of this book.

Here I think a little history is appropriate. A good starting point might be Thomas More's book "Utopia", where the residents of an imaginary idyllic island had a common culture, with all goods shared in common (much like life in a monastery) (More 1516). Later the Marquis Nicolas de Condorcet wrote a small book while in a French prison, awaiting execution, "*Esquisse d'un tableau historique des progrès de l'esprit humain*" ("Sketch for a Historical Picture of the Progress of the Human Spirit") published posthumously in 1795. In it, he envisioned a utopian future, including social insurance. It is considered one of the major texts of the Enlightenment and of historical thought, but he had no chance to elaborate the idea.

Condorcet's ideas were simultaneously taken up and elaborated by Thomas Paine, in the last chapter of his block-buster pamphlet, "*The Rights of Man*" (Paine 1791). Paine's book strongly supported the French Revolution, which is what it was mainly remembered for. But in the last chapter, Paine said that welfare for the poor is not charity, *but a natural and irrevocable right.* That view was radical at the time of the French Revolution. It is still radical, if a little less so, today.

Inequality is suddenly a hot topic among economists. Only a few years ago, it was regarded as a socio-political issue, not for mainstream economists (except Amartya Sen, Partha Dasgupta, and a few others). This changed when Nobel laureate Joe Stiglitz wrote "*The Price of Inequality*" (2012). Then, more recently, Thomas Piketty published his big (700 plus pages) book "*Capital in the Twenty-first Century*" (2014). It became a best-seller for a while when Paul Krugman gave it a rave review in his column in *The New York Times* and in the New York Review of Books, calling it "a magnificent, sweeping meditation on inequality" and touted it for a Nobel Prize in economics.

Piketty's core thesis is that low income and estate taxes on wealthy people has enabled them to invest their surplus (savings) in capital stock, resulting in growing disparities in ownership of wealth, especially in the form of land. He also points out that income is shifted from the poor to the rich, whenever economic growth is slow, because the rate of return on financial assets is larger than the rate of economic growth, and returns to labor. Land prices tend to increase even faster than population or economic growth (despite cyclic ups and downs) because the quantity of land in desirable places does not increase. Moreover, land value is the basic security behind virtually all housing and office mortgage loans. Hence the owners of land tend to become richer automatically, without doing anything to "improve" it.

Joseph Stiglitz has pointed out that wealth today is not necessarily related to capital in the traditional economic sense of the word (as "the means of production") (Stiglitz 2015, 2019). Most of the recent gains in financial

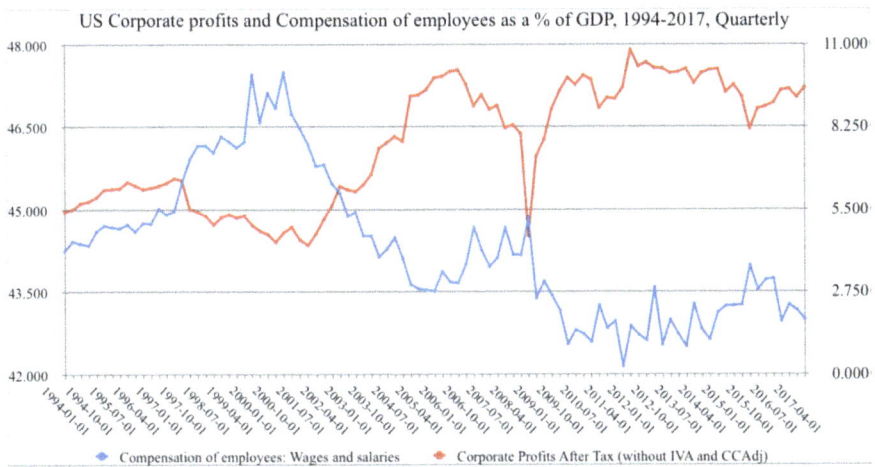

Fig. 1 Corporate profits and employee compensation

wealth, especially by the 1% at the top of the heap, arise from gains in unproductive asset values, especially of urban land and shares of large public companies. Much of those stock market gains arise from increasing corporate profits. The latter have risen, in turn, thanks to tax cuts, anti-union ("right to work") legislation, and "monopoly rents" due to unopposed consolidation (mergers and acquisitions) that reduce competition, and share buybacks. Corporate profits, as a share of GDP, are rising at the expense of employees' wages and salaries as Fig. 1 shows clearly.

You may ask: How is it that corporate executives have been able to raise profits so consistently year after year? In brief, it is because the counter-pressures that would have tilted the balance of power in favor of the workers in some other countries (notably Scandinavia) have been systematically weakened and destroyed in the US by legal means. In brief, laws and rules have been enacted by Republican politicians, elected in gerrymandered districts, in elections financed by special interests, to ensure that they stay in power. Stiglitz identifies the US version of capitalism as the "Reaganite ideology" of privatization: that private enterprise is "good" while government is "bad." Some Republicans even argue that tax cuts that reduce government revenues are desirable, because they are a way to "starve the beast" (ibid.).

Stiglitz correctly points out that this admiration for privatization is badly misplaced. He notes that the non-profit public education system in the USA—for all its faults—has produced the best system of universities in the world, whereas many of the for-profit universities (such as Trump University) have been scandalously bad. The primary problem for public education is (again) inequality. That inequality arises from local control and local financing. That is considered a virtue by some conservatives, but it also means that some school districts in wealthy districts are much richer than others in poor districts.

Consequently, the schools in wealthy neighborhoods can afford to pay teachers much better and hire better teachers. The public schools in Palo Alto California and Short Hills New Jersey (and other such places) are on a par with the best private schools, whereas public schools in Newark or Oakland are starving for funds. In Europe, by comparison, schools are funded by the central government and provide virtually equal educational quality in all schools, rather than depending on local real estate taxes. As a result, they do not have Harvard, MIT, or Cal Tech, but they have fewer people left out in the cold at the other end of the spectrum.

The US health system is by far the most expensive in the world, yet markedly inferior among the OECD countries. It creates millionaire specialists and administrators, while driving poor people with serious illness into bankruptcy. According to statistics compiled in 2012, a routine visit to the doctor's office cost about $30 in France or Canada ($36 in Chile, $11 in Spain) as compared to $176 in the USA. The cost of a hospital day in France in that year was $853 ($1,472 in Australia, $964 in Chile, $476 in Spain) compared to $12,537 in the USA.

The problem, again, seems to be privatization. Health-care corporations in the USA are created to maximize profits for shareholders, and they do this by trying to limit care to the healthiest people who need it least, while denying it to those in greatest need. They also have other gimmicks to maximize profits while passing costs on to the government (i.e. taxpayers). Needless to say, the shareholders tend to be among the wealthiest cohort of citizens, while the denizens of emergency rooms are among the poorest.

Recently my wife and I passed through the Panama Canal, surely one of the great engineering feats of all time. We learned that the private enterprise project to build the canal, by Ferdinand de Lesseps (of Suez fame), failed utterly in 1887. It destroyed the savings of many middle-class French investors and cost 6,000 lives of workers, due to diseases (malaria and yellow fever) they did not know how to treat or control. Only 13 years later, thanks to President Theodore (Teddy) Roosevelt, the Panama Canal project was taken over by the US government. The canal was built by the US Army Engineers (a government organization) below cost, in less time than expected, and with far fewer deaths from disease. So much for the inefficiency of government vs the private sector (Sekera 2016).

Back to the question posed at the beginning of this section: Is capitalism, as such, responsible for inequality? Early in April 2019, the House Committee on Financial Services held a Hearing, at which seven important US bankers testified. Most of the questions and answers were low-key, but a Republican Member of Congress, named Roger Williams from Texas, got headlines in *The Financial Times* by asking each of the seven bankers "Are you a socialist or are you a capitalist?" Of course, the question is unanswerable without caveats and definitions, and the bankers at the Hearing did not even try to provide serious or nuanced answers. But that question: "Are You a Socialist or a Capitalist?"—as though the difference between the two is black and white—is now floating in the air, so to speak. It may be more important in coming years.

The underlying weakness in the case for capitalism, in the sense of privatization of everything possible, is clear: It has a basic fault. Those with money—the wealthy—have all sorts of advantages, from superior education to better access to information, to better access to credit and access to political power. As Benjamin Franklin said, 250 years ago: "*Money can beget money and its offspring can beget more. The more there is of it, the more it produces every turning, so that the profits rise quicker and quicker.*" By allowing the "winners" to retain, build on, and to pass all or most of their winnings to their children (as Piketty points out), we get growing inequality. For long-term social stability, this fault needs to be addressed and compensated by other (political) means. American-style capitalism currently exaggerates this fundamental fault—employing most of the megaphones to drown out its critics—while failing to admit the existence of a crack in the foundations of the capitalist system. When structural foundations crack, the result is likely to be catastrophic.

This problem has been obvious to philosophers, social thinkers and as long as capitalism itself. But thanks to the Cold War and other international distractions, political argument in the USA has been focused on international power politics and civil rights, not on poverty and inequality as a continuing economic issue. Until the last few years, poverty and inequality have been seen in the Corridors of Power as minor perturbations of an otherwise perfect economic system. The self-satisfied attitudes of banking tycoons and their clients in corporate headquarters everywhere can be expressed, in the words of Jamie Dimon of J.P. Morgan in his 2018 letter to shareholders, in the phrase: "capitalism has lifted billions out of poverty." "Well," he adds (to an interviewer) "maybe the government should spend some more money on infrastructure and education. Maybe income taxes for the wealthy should be a bit higher." (He earned $30 million in 2018).

Never mentioned by Wall Streeters like Dimon et al., in their public pronouncements, is the fact that a significant fraction of their business—if not the major part of it—is lending depositor's money to people who borrow it, primarily to engage in "financial engineering." The game is marked by corporate mergers and takeovers, not by creating new enterprises or expanding existing ones. Every takeover makes a few wealthy people richer, and almost invariably leaves a lot of low paid workers without a job. A recent example illustrates the problem neatly. When Kraft (or the private equity fund behind it) absorbed Heinz—previously a family company—in 2012. The new owners, Warren Buffet's Berkshire Hathaway, with Brazil's 3G Capital, cut $1.5 billion in annual operating costs. This raised the profit margin significantly and enriched the billionaire shareholders (including the Heinz Foundation). This gain in profitability was achieved by cutting around 30,000 "working class" jobs at Kraft and Heinz.

Encouraged by this success, the owners of Kraft-Heinz looked for new targets. The one they chose was Unilever, a much larger firm. The financial engineers saw Unilever as a pot of gold waiting to be harvested (excuse the mixed metaphor). The proposed deal with Unilever was kicked off by an offer of $50 a share, representing an 18% premium over the then-current price of about $43 per share. It was generally supposed by the "market-watchers" that this was an opening bid that would need to go higher (to $60/share?) before succeeding. The takeover would have been financed by debt, secured by Kraft-Heinz, of course, and the debt would then become a liability of the "new" Unilever. To pay the interest on the new debt and double the profit margin, it would be required to cut costs much larger than those at Kraft-Heinz, probably not less than $3 billion per year from revenues of $53 billion. How many lost jobs? Without a sharp increase in revenues, I would guess at least 10,000 out of the global Unilever workforce of 168,000. To the "value-creating" capitalists, those are just numbers. To the employees and their families, they are something else.

This particular deal did not happen, partly because a certain deal-maker, named Michael Klein, called Warren Buffet and explained the "political sensitivities" of the likely job losses in the UK and the Netherlands. Also, Unilever's long-term record of returns to shareholders (9% p.a. for the last 20 years) is second only to that of Nestlé in the food industry. Hence, he pointed out, that Paul Polman, the CEO of Unilever, cannot be accused of incompetence or disregard of shareholders. Buffet pulled out, making him a hero.

But, as the *The Financial Times* noted, this case was exceptional: in the vast majority of such cases, the "activist" capitalists succeed in enriching themselves still further at the cost of employees, customers, and communities. The traditional finance industry is, in effect, "drilling for oil" (i.e. profits) on Wall Street, not on Main Street. Unfortunately, there are not very many "sleepy companies" with "lazy managers" left to squeeze. The rise in share prices in recent years was much faster than the underlying increase in corporate profits, that supposedly justified it. That makes it a bubble. The inevitable collapse may be very painful. The collapse of this financial bubble—like others—will increase inequality to still greater extremes. Will that be enough to induce politicians to make fundamental reforms? I wonder?

Again, the problem is that when the investments are profitable, the benefits are privatized by the bankers and their clients, the big borrowers, such as hedge funds and "private equity" firms. Those profits are not shared with the owners of the bank deposits backing the bank loans, unless those depositors are also investors in the stock market. But when things go wrong, as in 2007–2008, the costs are shared by those depositors, as well as by the taxpayers and all kinds of small businesses dependent on the bank. That fact alone explains why the bankers get richer than their depositors, and wealth tends to accumulate in a smaller and smaller fraction of the population.

Worse, much worse, the games they play with other people's money are essentially unproductive, contrary to popular ideology. It was not the banks that created new enterprises that lifted billions of people out of poverty. While the bankers like to claim credit for the success of our economic system as a whole, their ancient function of allocating capital to productive activities is now largely a fiction. The higher profits to shareholders of Kraft-Heinz have not created any jobs or financed anything productive. They are being invested in asset prices.

Today, banks allocate capital increasingly to activities that can only be called gambling. The fact that money moves around makes it look like ordinary economic activity, and it shows up as a contribution to GDP. But buying and selling financial instruments, whether stocks and bonds, or derivatives, or whole companies, is *not* putting money to work producing goods and services for society as a whole. The capitalism of "Wall Street" is a system that enables the wealthy to get richer at the expense of "Main Street," partly by exploiting the existing rules, and partly by manipulating the political process that makes and enforces those rules.

The list of mechanisms that are designed to privatize gambling gains while "socializing" the losses is not by any means complete. Most of these mechanisms have been created by changes in basic laws, promoted with good intentions, usually in favor of deregulation, so that "free enterprise" can function better with less government interference. These changes, such as the

Taft-Hartley Act that weakened trade unions, the Securities and Exchange Commission (SEC) decision to allow investment banks to incorporate themselves, the 1982 action by the SEC Chairman, opening the door for corporations to buy their own shares (previously illegal), the 1998 law voiding the 1933 Glass–Steagall banking law, or the 2016 Tax law cutting corporate income taxes financed by enormous increases in the national debt, can also be reversed by lawmakers. They should be.

Fixing Capitalism: Is It Time for UBI

I pointed out at the beginning of this book that if the game of hyper-competitive "winner-take-all" capitalism goes too far—leaving too few winners and too many losers, there will be a revolution. The revolution will almost certainly be violent, and the wrong people—"strong men," like Mussolini and Hitler—will come to power in messy situations.

The way to keep the game going is to put more money into the system. This can be done in several ways. One way is to tax the rich as many social democrats will advocate. The problem is that it is both unpopular and ineffective. Maybe the unpopularity can be reduced by a clever public relations campaign, arguing that the billionaires of today did not "earn" their vast fortunes (as most of them firmly believe) but are more like "lottery winners" who happened to be in the right place at the right time. Perhaps a Gandhi-like "voice of the people" can persuade them that they have a moral duty to pay for the environmental damage they have caused—being the major beneficiaries of the world's economic wealth. I think it is a good argument, but I do not think it will persuade many billionaires.

The next option for getting more money into the game is to borrow and/or "print money" (economists call it "quantitative easing" or QE) and hope that the increased consumption resulting from lower interest rates will generate enough economic growth to compensate for the inflationary effect. This gimmick worked briefly in the USA, for a few years, even though the banks that got the money did not use it to increase lending to small businesses. Instead the benefits of QE went to the hedge funds and private equity funds, so most of the benefits went to the rich after all.

The US also "benefited" from a $1.5 trillion tax cut in 2017 for corporations and the rich, paid for by increasing the national debt. These two interventions have kept the US economy from going into recession, temporarily. The rest of the world is already in trouble, and the high tariffs of the Trump administration are making it worse. The trade war with China is not likely to be resolved any time soon, though Trump will claim some sort of victory, letting him cut tariffs in time for the next election. But none of this manipulation will solve any of the deeper problems.

It is time to consider another way of getting money into the system, without funneling it directly through the banks to the wealthy. It is an old idea, called the negative income tax (Friedman) or the Universal Basic Income (UBI). Among the many versions of redistributive tax that have been proposed in the past, one that keeps popping up in the discussion is the single tax (*impot unique*) on land. Such a tax has been proposed by Baruch Spinoza, by John Locke, in the seventeenth century, by the French physiocrats in the eighteenth, and more recently by Henry George in the nineteenth century. Henry George's basic argument is that wealth produced by work should be owned by the worker, whereas wealth from land ownership should be allocated equally to all, as a source of universal income. His tax on land was proportional to the area of land involved (i.e. wealth), not income.

The modern version of the UBI started with Henry George and his radical "single tax" proposal of 1879. Mabel and Dennis Milner, a Quaker couple active in the anti-war movement of 1914–1918, took up the idea. They wrote a pamphlet that later became an academic paper entitled "A scheme for a state bonus" (Milner and Milner 1919). They argued for a weekly income, not subject to any conditions, based on the citizen's "moral right" to the means of subsistence, irrespective of deserving. This echoed Tom Paine's view.

Since the Milners wrote their piece, the idea has been proposed in various forms by quite a few others, including Bertrand Russell, George Bernard Shaw, Milton Friedman (the "negative income tax"), John Kenneth Galbraith, and—surprise!—Richard Nixon. Apparently in the late 1960s, it was widely agreed that the some UBI system would soon be adopted in the USA and Canada. In fact, Nixon supported it and twice got it through the House of Representatives. It was blocked on the floor of the Senate by Democrats, who wanted a more generous version than Nixon was offering, and thought that opposing the plan from the House would force another debate and another vote. That was a strategic error, to put it mildly.

Ironically, Seymour Hersh (the reporter who tracked down Lt. Calley, the My Lai story, and the Abu Ghraib atrocities) was press secretary for Eugene McCarthy's primary campaign in 1967. One of the suggestions he and others presented to McCarthy was a universal basic income. Apparently Wilbur Cohen, secretary of health, education, and welfare in the Johnson administration had done some research on and found feasible. McCarthy paid no attention; too bad—it might have got him elected.

Since 2016, the idea of a universal basic income has "gone viral" and for good reasons. The reasons have been summarized very succinctly in a "Manifesto" by Ray Dalio, the billionaire founder of Bridgewater Associates, the largest hedge fund. Dalio's personal net worth has been estimated as $17 billion, which is relevant only because he has a lot of experience in the world of money. Here are a few of the main points, taken from his manifesto—circulated on the Internet: "Why and How Capitalism needs to be reformed." He breaks the economy into two parts, the top 40% and the bottom 60%. He shows that prime-age workers in the bottom 60% have had no real income growth (adjusted for inflation) since 1980.

In my metaphor of the economy as a poker game, that 60% group is "out of the game" (actually they were never in the game). They have no surplus to invest. They cannot "play" in the game of wealth accumulation. Most of them are poor and will stay poor. According to an Federal Reserve Board (FRB) study, two-thirds of that bottom group (the lowest 40% of the US population) would be unable, or have great difficulty, to raise $400 in case of an emergency. Only a third of that 60% group (20% of the total population) have any savings at all. What savings they have is mostly home equity. Quite a bit of that was wiped out by the crash of the real-estate price bubble in 2008, which wiped out over $5 trillion in home equity.

But since 1980, the incomes of the top 10% have doubled and the income of the top 1% has tripled. Another sign of increasing inequality is that the chances of escaping from that bottom pool have declined. In the year 1970, most (90%) children grew up to earn more than their parents. In 1990, the odds of moving up from the bottom quintile (0–20%) to the middle quintile (40–60%) in 10 years were 23%. By 2011, they were down to 14%. This is a sharp decline in "upward mobility."

In the late nineteenth century, there was a myth promoted by popular dime novels for boys, by Horatio Alger Jr. who was the best-selling author of his time (over 100 books published). They conveyed the message that anyone, no matter how poor, can become "a self-made (rich) man" by hard work. If anyone, it was Alger who articulated the "American Dream." There are just enough real-life examples—George Soros, Larry Ellison, Roman Abramovich, Leonardo de Caprio, Howard Shultz, Oprah Winfrey, Sheldon Adelson, and some you have never heard of—to "confirm" that thesis. But, in 1970, boys still had a 90% chance of doing better than dad. By 2017, that 90% was down to 50% and falling. (There are no comparable data for girls). Sure, there are still some celebrated examples of golden success. But it was a lot easier when they did it, than it is now. Today, only 8% of the people in the top quartile (the top 25%) had a father in the bottom quartile.

Ray Dalio's manifesto on why American capitalism is not working well goes on to explain the underlying roots of its failure. He blames education first of all. After presenting a truckload of statistics, he says *"Though there are bright spots in the American education system, such as our few great universities, the US population, as a whole, scores very poorly relative to the rest of the developed world in standardized tests for a given education level. … The US is currently around the bottom 15th percentile in the developed world."* For a country that brags about its wealth and accomplishments, that statistic is shocking. Dalio has a lot more to say about the American health-care system, because the bad health care available to poor children and the bad teachers available in poor school districts have a deleterious effect on those children's learning ability, not to mention their later behavior as adults.

Needless to say, I think the Dalio manifesto is, by itself, a powerful argument for the Universal Basic Income proposition. But he himself has not supported the UBI (yet) and one must wonder why. For the moment, I assume that he would rather spend the money on targeted educational and health-care programs. That alternative is the most seductive argument against the UBI proposition. I will come to it, in a few paragraphs.

The debate about UBI has revealed two distinct groups of backers of the idea. One new group consists of extremely wealthy people—some of them Silicon Valley plutocrats, such as Richard Branson, Nick Hanauer, Elon Musk, and Mark Zuckerberg, to name five—who have an economic stake in the continued development of digital technology, automation, robotics, and artificial intelligence. They are well aware that the likely consequence of further technological progress—if that is the right word—will be the disappearance of large numbers of traditional jobs. The socioeconomic consequences of such massive job losses are scary, and not least to those who will be blamed for it when the time comes.

The other group of backers consists of people who acknowledge the unemployment issue, but think the problem is much deeper. Mabel and Dennis Milner, a Quaker couple active in the anti-war movement in 1914, wrote a pamphlet entitled "A scheme for a state bonus," echoing Tom Paine's view (Milner and Milner 1919). It later became an academic paper. They argued for a weekly income, not subject to any conditions, based on the citizen's "moral right" to the means of subsistence, irrespective of deserving.

Since the Milners, the idea has been proposed in various forms by quite a few others, including Bertrand Russell, Milton Friedman (the "negative income tax"), John Kenneth Galbraith, and—surprise!—Richard Nixon. Apparently in the late 1960s, it was widely agreed that a UBI system would soon be adopted in the USA and Canada. In fact, Nixon (a Quaker himself) supported it in the 1970s and twice got it through the House of Representatives. It was blocked on the floor of the Senate by Democrats who wanted a more generous version than Nixon was offering. It seems they thought that opposing the plan from the House of Representatives would force another debate in the Senate and another vote. That was an historic strategic error.

The most popular of the recent books on UBI include the following: *Raising the Floor: How a Universal Basic Income Can Renew Our Economy and Rebuild the American Dream* (Stern & Kravitz 2016); *Basic Income* (van Parijs 2017 & Vanderborght); *Utopia for Realists* (Bregman 2017) *Basic Income: A Guide for the Open Minded* (Standing 2017); *Give People Money: How a Universal Basic Income would End Poverty* (Lowrey 2018).

Guy Standing, who has been plowing this field for 30 years, says that there are three basic reasons for embarking on a UBI program. Reason (1) is what he calls "social justice." This means recognition *that everyone living now, no matter their degree of personal economic success, owes most of it to those who came before.* From that perspective, the poor deserve to be provided with a minimum means of living, as a "dividend" from that accumulation of wealth from the past. Today's billionaires did not create most of that wealth, they only acquired it.

Reason (2) is that people—regardless of income—deserve to make their own decisions in regard to money, free of arbitrary conditionality and coercion from hierarchical "superiors" or faceless functionaries. I think most conservatives will agree. But they, as a group, are still convinced that the poor are undeserving, and that poverty is a character problem. They are wrong about that. The current debate is also exposing how deeply rooted, yet rotten, is the still-dominant Protestant Ethic, which attaches moral virtue to work, and hence to wealth, while dismissing poverty as a "personality defect" (in the deathless words of Margaret Thatcher).[1]

Here is what she actually said: *"Nowadays there really is no primary poverty left in this country. In Western countries we are left with the problems which aren't poverty. All right, there may be poverty because people don't know how to budget,*

---

[1] I can't find the actual source, but the quote has been re-quoted many times and widely discussed, see Google.

don't know how to spend their earnings, but now you are left with the really hard fundamental character—personality defect."

The truth is otherwise: it is poverty that causes bad behavior. Her view of the world is rapidly and justifiably losing credibility, even to conservatives. Harvard economist Sandhil Mullainathan and Princeton psychologist Eldar Shafir argue that it is indeed a matter of willpower and bad decisions (Mullainathan and Safir 2013). But Thatcher et al. have it backward. It's not that foolish choices make you poor; it is that poverty's effects on the mind lead to bad choices. Living with too little imposes huge psychic costs, reducing our mental bandwidth and distorting our decision-making in ways that dig us deeper into a bad situation. Reason (3) is the need to provide basic security, as a human right, harking back to Thomas Paine and the Milners. I agree with that argument, notwithstanding the fact that some undeserving people will be able to waste their time and their talents. That is the price of progress. I also think there is a reason (4): Call it National Security. I think UBI is the only practical way of reversing the current trend toward extreme economic inequality, which is the subject of this book. If the current trend is not reversed, and soon, the social and political consequences will be very bad. Global nuclear war is not impossible.

Objections and Counter-Arguments

The classical objections to a guaranteed basic income, voiced in a variety of different, but roughly equivalent ways, are that the recipients will stop working and become lazy "couch potatoes" watching game shows or sports on TV all day. Or they will become druggies or worse. Who will do the dirty jobs like cleaning the public toilets or harvesting the tomatoes? This fear has been expressed many times, often in harsh language, by wealthy conservatives or their acolytes—people lacking any actual evidence to support their assertions.

According to numerous survey results, there is more unpaid labor in the world than paid work. Women do most of it, especially in developing countries, and child care is the major category. Single mothers with young children probably work harder with little or no reward, than any hedge fund manager or banker ever worked. Even begging in the street is hard work. The alleged connection between hard work and financial success, if it ever existed, is now invisible. In this day and age, there is no such connection.

There is another quasi-moral objection. Economists use the term "free riders" in reference to people who obtain monetary benefits like subsidies, without paying for them. For instance workers, who benefit from union contracts, without paying union dues, are "free riders." Another example is the "tragedy of the commons" which refers to situations where public goods, such as fisheries, are overused to the point of collapse (Hardin 1968). (That outcome actually happened to the New England cod fishery in the 1970s.) Economists devise incentive schemes, such as the "Coase theorem" (mutual agreement among beneficiaries to share costs) to prevent free-riding. But this is possible only in the theoretical case of zero transaction costs (Coase 1960).

Yet when people are polled on what they themselves would do if they did not have to work in order to eat, they overwhelmingly answer that they themselves would not quit working, but that, if free to do so, they would want a different kind of work. Almost all agree that, when given the opportunity to live at a bare subsistence level, without working, very few would choose to do so. In fact, there have been many experimental tests of the question in a number of countries (Bregman 2017). But the experiments show almost no evidence of the lazy, slothful, "free rider" behavior expected by conservatives. What we do know from the evidence of many experiments in a number of countries is that crime goes down among the group being studied, while health improves and school performance of children also improves.

One counter-example is the case of an airline clerk in New York in the 1950s, who wanted to be a writer. She wrote a few things, not very successful, but an agent took her on. After some more disappointments, some friends banded together to give her a gift of a year's wages with a note: "*You have one year off from your job to write whatever you please. Merry Christmas.*" The book she wrote (after that year, and another) was called "To Kill a Mockingbird" and her name is Harper Lee (actually Nelle Harper Lee). That book won a Pulitzer Prize among other honors and has sold more than 30 million copies.

Granted, there are a lot more would-be writers, would-be athletes, would-be artists, and would-be actors than there are openings. Many will be disappointed. But the UBI would certainly allow some to succeed. The amount of art in the world would also increase.

So much for the "free rider" issue. Social justice is another issue. The billionaires in Silicon Valley (and Manhattan) are all convinced that their earnings are proper rewards for their contributions to society. They need to be reminded that they are essentially lottery winners, like the property-owning Grosvenors of London and the Stuyvesants of New York. Some of them are honest enough to acknowledge the fact that they were in the right place at a lucky moment, their virtue was to be smart enough to recognize, and lucky enough to have the means to invest in a good opportunity when it came along. But they all stand on the shoulders of thousands of others who received little or nothing in reward for their efforts.

Ayn Rand's image of "Atlas" supporting the world on his shoulders is exactly right, except that "Atlas," in reality is not Bill Gates or Steve Jobs or Warren Buffet or their counterparts. Atlas is the heritage of physical capital, infrastructure, education, and knowledge that present-day entrepreneurs can build upon. Bill Gates invented nothing. MS-DOS was a program he bought. Steve Jobs invented nothing. He was just smart enough to exploit the inventions of people at Xerox's Palo Alto Research Center (PARC). Warren Buffet did not invent anything. The hedge fund guys do not invent anything except tricky ways of tax avoidance.[2]

The unfortunate fact is that the true inventors and innovators (mostly) do not get rich. They do not know how. They lack the skills and resources. The owners of major wealth today—the winners of the capitalist game—are the (mostly) men who have mastered the tricks of making money from borrowed money, by means of the 3Ms: manipulation, management, and maximization. But today's owners of wealth did not "earn" more than a tiny fraction of the world's financial wealth, never mind the natural wealth of the environment that they exploit (and destroy). In justice, they should not have it. Their claims to do compensating social good by setting up philanthropic foundations to promote various high-sounding purposes, such as medical research, are nothing more than cop-outs for "original sin." By sin, I mean, the enjoyment of undeserved benefits acquired by doubtful means from the contributions of many other people. My point is that supporting UBI would be a more appropriate compensation.

There are several good reasons for taxing the very rich, if it can be done effectively (another question). Next on the list (after social justice) may be the fact that they—the 1% and the 0.1%—simply have too much of the wealth that exists now, leaving too little for the rest. It is not only deeply unfair; it is much worse than that, it is very dangerous, not only for them. As venture capitalist Nick Hanauer has been saying to his fellow plutocrats on U-tube— "the pitchforks are coming" and when they do come, the destruction will be terrible. The French and Russian Revolutions and the Communist takeover in China illustrate what can happen once the avalanche gets started.

Note that I am talking about the mal-distribution of existing wealth, not of income. Wealth is mostly inherited as Piketty points out (Piketty 2014). It is also largely—except for collectibles and works of art—the same as capital stock. It is thus closely correlated to the means of production. It is important

---

[2] I do not include Elon Musk or Steve Bezos in the group of undeserving rich. They are truly wealth creators, although they, too, are standing on shoulders of others.

to recognize that the total wealth of the world is limited at any one time. Whatever capital the top 1% owns and controls is—by definition—not available to the rest of us. There is no question of a "rising tide" of wealth that benefits everybody. It is more nearly a binary situation: the resources owned by the plutocracy are unavailable to the rest of society (us). That is why the poor people at the bottom of the pyramid are out of the capitalist game. They have no savings and nothing to invest, which is why they—and their children—have virtually no chance of escaping from poverty, no matter how smart they are or how hard they work.

Income inequality is probably a little less dangerous than wealth inequality, in terms of the coming of the pitchforks. But there is quite a lot of overlap. Top executives being paid large amounts of money—I will not say "earning it" because I do not believe they do earn it—soon become plutocrats themselves. Moreover, when the top executives of a firm take an outsize share of gross profits for themselves, there is less for the rest of the employees. That is undeniable.

The argument in favor of large pay packages for CEOs rests largely on the fact that corporate profits are less and less based on sales, product quality and productive efficiency, and more and more based on lobbying and "financial engineering." It is why top executives are increasingly being recruited from consultancies (like McKinsey), "private equity" groups or from other companies, not from within. This is not the place for an extended discussion of what "financial engineering" means, except that it has nothing to do with real engineering. Most readers who have got this far know what it means, and I shall leave it there.

The same logic sits behind mineral resource rent taxes—such as the first incarnation of the now-abolished mining tax. When the international price of a resource goes up, those who own the resource (every Australian) receive little benefit. The benefit goes to the mining companies even though they have done nothing to facilitate those price rises and they do not own the material whose price has risen. This is unearned income and could be taxed in order to return the income flows to the public.

Most businesses in Australia would greatly benefit from a tax shift to economic rents with a commensurate reduction in company tax and the abolition of inefficient taxes such as stamp duties and insurance taxes.

Vast sums of money that are currently directed toward rent-seeking would be redirected into productive activity, generating employment and diversifying the economy. Boom and bust property cycles would be flattened due to reduced speculation and, as a result, the broader scale ups and downs of the business cycle would be somewhat moderated.

The political hurdles to serious tax reform are very high. However, the consequences of not reforming the tax system are severe. Tax reform policies are easy prey for opportunistic political opponents. This is why we need some clear principles for tax reform that are clearly explained to the public.

Liberal politicians should favor shifting taxes off productive business and onto economic rents and the exploitation of shared resources because such reforms target market failure and free up productive and sustainable businesses to flourish. Labor politicians too should approve of these principles because they reduce taxes on labor and shift them onto the rent seekers who contribute little to society. The inherently progressive nature of most rent taxes should also appeal to The Greens, the Labor left, and the increasing number of others concerned about economic inequality.

Our politicians will need courage to stand up to powerful individuals and groups who have an interest in maintaining the status quo. They can get that courage from the rest of us who stand to benefit from a taxation system that supports a more productive and sustainable economy.

Paying for UBI

Now it is time to come to the big problem: how to finance the UBI. Most people, starting with Henry George and the current generation of "Georgists," as well as Bernie Sanders Elizabeth Warren and most liberals, advocate progressive income taxes and/or wealth taxes on the rich. However, the most popular argument for using taxes to finance UBI is probably the worst one, namely that "they shouldn't be so rich in the first place, they don't deserve it, so we have a moral right take it to take it away, or worse, the moral right to punish them for being so rich". That motivation was probably behind the "reign of terror" in France in 1793–94.

Whether the last sentence is accurate or not, it makes a point. Redistribution by taxation is unavoidably linked to tax reform, which is very difficult and complicated subject. The political hurdles to serious tax reform are very high, but the consequences of not reforming the tax system are probably higher. Tax reform policies are easy targets for opportunistic politicians. Hence, the need for clear principles for tax reform that can be clearly explained to the public. That task is sadly beyond my ability.

I start with the axiom that taxes are necessary to finance the government, which means paying for basic government services. There are major differences among people with different ideas about government, as to what services need to be financed, and at what level. The government budget at every level in every country also includes large interest payments on existing debt. As the TV pundits tell us every night, US government debt has been rising for quite a long time—roughly since 1970—thanks to persistent budget and trade deficits, thanks (in turn) to tax cuts for the rich. International trade is up. Corporate profits are up. Yet middle-class wages and standard of living have *not* been rising during the last 40 years. The lowest income group has lost ground and become poorer than it was 40 years ago. This is probably the major single cause of voter dissatisfaction with neo-liberalism and globalization.

Up to now, government debt has been rising as debt service at the Federal level (interest payments) has been simultaneously declining. This is because interest rates have been reduced after every crisis by the Federal Reserve to stimulate the economy. That pattern changed in 2017. Since then, interest rates on government debt (bonds) have started rising. This is because returns on long-term (10 years) bonds have started to rise. That means debt service will get more and more expensive in coming years unless government spending is cut drastically.

In the USA, military expenditure consumes a large fraction of total revenues, and few American politicians dare to question the size of that budget, even though it is far larger than most other countries in the world and there is no existential threat. But the US spends far less than other countries on some other services, such as infrastructure, education, health, and housing. This is clearly not the place to address those differences and inconsistencies. My proposal below makes no attempt to adjust for real-world constraints. However, I cannot avoid some comments on taxation, per se.

First of all, there is a critical difference between earned and unearned income for tax purposes. Earned income is payment for useful work of some kind, while unearned income is called "rent" and people with unearned income are "rentiers" in econo-speak. Most economists favor taxes on rent, or rent-seeking, rather than on earned income because taxing rent reduces inequality more effectively than taxing labor. Examples of rent taxes are capital gains taxes, real estate taxes on land values, carbon taxes, resource extraction taxes, and taxes on the use of the electromagnetic spectrum. (Excise taxes on gambling, tobacco, and alcohol are not rent taxes.) Most democratic countries currently get the bulk of their revenues by taxing labor, not rent, i.e., personal income taxes and corporate income taxes. This is not because taxes on labor are more efficient, or more fair and just than taxes on rent. It is mostly because they are easier to collect.

Taxes on wealth, per se, are another kettle of fish. It has been pointed out that a wealth tax serves as "negative reinforcement," in the sense that it motivates the productive use of assets. A wealth tax also taxes capital that is not productively employed. Thus, a wealth tax can be viewed as a tax on potential income from unemployed capital. A net wealth tax—even if it is quite small—can be treated as an income tax on potential (imputed) income. Hence the capital gains, estate, and gift taxes can be eliminated (if the financial part of the wealth stays to be taxed). Guy Standing has advocated a sovereign wealth fund, financed from levies on rentier income, not only from undeveloped land but from all kinds of assets, both physical and financial (Standing 2017).

But financial wealth is very slippery, and it moves very easily to "safe havens"—mainly island countries with low or zero tax rates set up to attract wealthy clients. To eliminate such havens would require action at an international level, taking years or decades to negotiate. For this reason, it is not a serious candidate at present, despite rhetoric from liberal Democrats such as Bernie Sanders and Elizabeth Warren.

On November 9, 1999, in that same year of the "Dot Com" stock market boom, Donald Trump—now President Trump—proposed (on CNN) a one-time wealth tax of 14.25%. The idea was to raise $5.7 trillion and wipe out the US national debt. Presumable much of that 14.25% tax would have mostly been covered by the stock market gains up to that point. (Do not bother doing the arithmetic. Donald Trump does not.) Unfortunately the market peaked a few months later and the taxable gains disappeared without being captured for productive use.

As regards the notion that high taxes reduce growth, it is simply false. Top income tax rates in the USA and the UK during and long after WW II were around 82% in the USA and higher in the UK, far above current rates. The top rates for US income taxes were still high (70%) until the Reagan administration cut them drastically. In the past income taxes were higher, and so was US economic growth. That argument does not compute.

You have been waiting for me to get to the point. So here is my—hopefully feasible—proposal. Assume the US GDP of $20 trillion, slightly more than the current level. Government expenditures in the USA, at all levels amount to about 40% or $8 trillion, of which $4 trillion is from the federal government. This includes something like $500 million for debt service. Currently individual income taxes (on labor) account for about half (51%) of all US federal government revenue, with payroll (social security) taxes from employers' accounting for another 35%. Corporate income taxes now account for a mere 9%, down from 25% or so 50 years ago. Excise taxes are now very small fraction of the total. Gasoline taxes, plus taxes on tobacco and alcoholic beverages, all together account for only 8% of US federal government revenue.

Suppose that UBI means everyone over the age of 18 receives a monthly "social dividend" of $1000, deposited in a bank. I assume that UBI would cost around $3 trillion per annum more than current federal tax revenues, but it would also replace a part of the existing SS retirement program, paid for by payroll taxes ($1.3 trillion p.a.) I guess that the net extra cost of UBI, after adjustments, would be around $2 trillion p.a. This would raise the total costs of the Federal government to $6 trillion p.a. without taking into account productivity changes.

Most liberals stop thinking creatively at this point and assume that the extra cost of UBI must be paid for by taxing the rich. Since that seems to be politically impossible, many liberals regard UBI as a pipe dream, not a serious political proposition. However, there are other possible sources of funding. I suggest that the labor share of US government revenue, now 86%, could be reduced to 80% in the near term and perhaps 70% in the long term, while the excise share, now a mere 8%, could easily be increased to 15% or more (or a larger total) by cutting income taxes and adding new taxes on carbon emissions and other things. That 15% would pay for around $1 trillion of the assumed $6 trillion total of federal government expenditures (including UBI) or half of the additional cost.

I would start by imposing a new excise tax on carbon emissions, to be paid by primary producers and importers of hydrocarbons and products with "embodied" carbon (like plastics or Portland cement). The excise tax rates could be set to provide annual government revenue of $1 trillion, mostly from carbon emissions, while also cutting those emissions by at least 10% (Maybe more). This could bring motor fuel costs in the USA up to European levels. The excise tax rate would have to rise as emissions decline to keep the revenue stream more or less constant.

No doubt the higher excise taxes on fuel would be resisted fiercely, as "regressive," if introduced on its own. But as part of a package, where most drivers also receive a supplementary income (the UBI). most low-paid workers would be better off, not worse off. Only the rich, with fleets of limousines and private jets, would pay more.

The negative reinforcement effect of carbon (exergy) taxes would encourage the search for non-carbon alternatives (and the search for more efficient means of recovery and recycling rare metals). Prices of some materials and consumer products would thus increase somewhat, of course, but not as much as the compensating savings from using more renewables, notably wind and solar energy. The price increases would mainly affect the well-to do, with several homes, expensive cars, private planes, etc. People using public transportation and not owning cars would pay very little. The overall effect of the package would be progressive, not regressive.

The negative reinforcement effect of higher carbon (exergy) taxes—or "scarce resource taxes"—would encourage the search for non-carbon alternatives and the search for more efficient means of recovery and recycling rare metals. Prices of some materials and consumer products would thus increase somewhat, of course, but not as much as the compensating savings from using more renewables, notably wind and solar energy.

There is theoretical support from several recent econometric studies that show that useful energy (exergy = work) is extremely underpriced, compared to human labor.[3] What this means is that the optimal (profit-maximizing) combination of input factors for the economy would use more exergy and less labor than it does now. That means automating everything possible. If, for social and environmental reasons, society needs more jobs and fewer carbon emissions, the way to achieve that is to increase the tax on energy (carbon) while reducing taxes on labor. The UBI—as a package—is therefore a way of un-taxing labor, in effect, by supplementing worker's personal incomes.

I propose another excise tax: an electromagnetic frequency spectrum tax, or internet tax. It should also be imposed, set to bring in annual revenues of the order of $1 trillion. When a tele-communication company uses a part of the electromagnetic spectrum—a frequency band—for profit making, nobody else can use that frequency band. The auctioning of rights to use the electromagnetic spectrum is an effective type of economic rent tax. The spectrum gets put to efficient use and the public is compensated for giving up a shared resource. The company then profits according to how well they use the resource rather than simply because they have a monopoly on use of it. There is bipartisan support for the auctioning of electromagnetic spectrum, but the principle can be applied much more broadly. The Internet is a public resource, and use of it needs to be allocated fairly by charging realistic prices for its use.

This tax could be collected partly as TV access fees—already common in Europe—fees on business data transfers and fees on Internet usage paid by mass message senders (like postage on first-class letters). Frivolous marketing use of the Internet, such as unwanted "junk mail"—much of it fraudulent—would be strongly discouraged by such a tax. Even "big data" users based on private servers would not be affected, while the tax would reduce internet congestion resulting from underpriced public uses.

---

[3] The details are technical but the main point is that it makes sense to make labor cheaper (by taxing labor less) and make resource inputs (especially fossil energy) more expensive by taxing resource inputs at higher rates. The well-known "cost share theorem"—which says that the productivity of any factor is equal to its cost share in the national accounts—is not applicable. It applies to an ultra-simplified model of two factors in equilibrium, and subject to other assumptions that are not valid. For proofs of this statement, see (Ayres 2000, Kuemmel et al. 2008, Kuemmel, Ayres, and Lindenberger 2010).

Third, I would introduce a value added tax (VAT), in the USA, explicitly to cut and replace personal income taxes for the middle class. This would be effectively similar to a tax on consumption. As with the carbon tax, it would be felt least by low-income people and most by big spenders.

The economic impact of the UBI together with the three tax changes in the USA would be roughly as follows. First, bank deposits from UBI would increase significantly, perhaps by the whole $2 trillion. Experience says that low-income beneficiaries would first pay off their high-cost credit card loans and student loans (though maybe not all at once). Money left over after that would be spent on household goods and services, albeit some of those goods and services would cost more thanks to indirect carbon and EM spectrum taxes. Private debt would fall, but overall the government debt might increase equally, perhaps by $500 billion p.a. The money supply might increase from 2% to (maximum) 3% p.a., a level central banks have been trying for years to achieve without success. N.B. the UBI would have a stimulus effect on economic growth comparable to QE, but not utilizing the banks for resource allocation.

As noted, UBI would actually cut some existing government costs, both for targeted welfare services that would become redundant, and even for prisons and police. Higher personal incomes available to spend on goods and services would also generate more tax revenues for the government. It is unclear how much would be added to the current intake, probably less than the net cost of the UBI. But the net deficit at the end of the day might be quite small or even non-existent.

My rough analysis above is only a first approximation of the costs and benefits of UBI ("helicopter money"). It would need to be extended and no doubt revised in various ways by the Congressional Budget Office (CBO) and numerous private sector interest groups, before being considered seriously by policy-makers. But the main point is to present the changes as a package, not as a series of individual change, most of which would be individually shot down by special interest groups.

So why not get started with the detailed analysis?

References

Ahamed, L. (2009). *Lords of finance: The bankers who broke the world*. London: Penguin.
Allenby, B. (2002). Observations on the philosophical implications of earth systems engineering and management. In *Batten Institute Working Paper*. Charlottsville, VA: Batten Institute.
Arthur, W. B. (1988). Self-reinforcing mechanisms in economics. In P. W. Anderson, K. J. Arrow, & D. Pines (Eds.), *The economy as an evolving complex system* (pp. 9–32). Redwood City, CA: Addison-Wesley Publishing Company.
Axelrod, R. (1984). *The evolution of cooperation*. New York: Basic Books.
Axelrod, R., & Dion, D. (1988, December). The further evolution of cooperation. *Science, 242*, 1385–1390.
Ayres, R. & Olenick, M. (2017). Fontainebleau.
Ayres, R. U. (2014). *The bubble economy: Is sustainable growth possible?* Boston, MA: MIT Press.
Ayres, R. U. (2014). The economic growth enigma: Capital, labor and useful energy. *Energy Policy, 64*, 16–28.
Bagehot, W. (1999). *Lombard Street: A description of the money market*. New York: Wiley. Edited by The Economist, London.
Bagehot, W. (1873, 1999). Lombard street. *The Economist*. London: The Economist.
Barry, R. (2014). *Crisis and complexity*. Vol. 1: First Principles Pty.
Black, F., Jensen, M. C., & Scholes, M. (1972). The capital asset pricing model: Some empirical tests. In M. C. Jensen (Ed.), *Studies in the theory of capital markets* (pp. 79–121). NYC: Praeger.
Black, F., & Scholes, M. (1973). The pricing of options and corporate liabilities. *Journal of Political Economy, 81*(3), 637–654.
Boehm, C. (1999). *Hierarchy in the forest: Egalitarianism and the evolution of human altruism*. Cambridge, MA: Harvard University Press.

Bregman, R. (2017). *Utopia for realists*. London: Little Brown & Co/Hachette Book Group.
Bresson, A. (2005). The origin of Lydian and Greek coinage: Cost and quantity. In: *3rd International Conference of Ancient History*. Shanghai: Fudan University.
Buffet, W. (1984). The superinvestors of Graham-and-Doddsville. *Hermes*, 4–15.
Burns, A. F., & Mitchell, W. C. (1946). *Measuring business cycles*. National Bureau of Economic Research (NBER): New York.
Carpenter, N., Jr. (1916). The Westinghouse Electric and Manufacturing Co. The General Electric Co, and the Panic of 1907 (part II). *Journal of Political Economy*, *24*(4), 382–399.
Coase, R. (1960, October). The problem of social costs. *Journal of Law and Economics, 3*, 1–44.
Colombo, J. (2012). *The British "Railway Mania" bubble*.
Cummings, J. B. (2006). *A brief Florida real estate history*. Tampa, Petersburg: Appraisal Institute, Region X.
Dalio, R. (2019). *Why and how capitalism needs to be reformed*. New York: Bridgewater Associates.
Diamond, J. (1998). *Guns, germs and steel*. New York: Vintage Books.
Dosi, G., Freeman, C., Nelson, R., Silverberg, G., & Soete, L. (Eds.). (1988). *Technical change and economic theory, IFIAS research series number 6*. New York: Francis Pinter.
Ellis, C. D. (2009). *The partnership: The making of Goldman-Sachs*. New York: Penguin.
Fama, E. F. (1970). Efficient capital markets: A review of theory and empirical work. *Journal of Finance, 25*(May), 383–417.
Forrester, J. W., Graham, A. K., Senge, P. M., & Sherman, J. D. (1985). An integrated approach to the economic long wave. In: *Long waves, depression and innovation, Siena/Florence*.
Freeman, C. (Ed.). (1996). *Long wave theory*. Cheltenham, UK: Edward Elgar.
Friedman, M. (1962). *Capitalism and freedom*. Chicago: University of Chicago Press.
Friedman, M. (1970). The social responsibility of business is to increase its profits. *New York Times Magazine, 32*, 122–126. 13 September.
Galbraith, J. K. (1954). *The great crash 1929*. Boston: Houghton Mifflin.
George, H. (1879). *Progress and Poverty: Increase of want with increase of wealth*. New York: D. Appleton & Co..
Goodwin, R. M. (1967). A growth cycle. In C. H. Feinstein (Ed.), *Socialism, capitalism and economic growth*. Cambridge, UK: Cambridge University Press.
Goodwin, R. M. (1987). A growth cycle. In C. H. Feinstein (Ed.), *Socialism, capitalism and economic growth*. London: Cambridge University Press.
Graeber, D. (2011). *Debt: The first 5000 years*. New York: Melville House Publishing.
Hardin, G. (1968). The tragedy of the commons. *Science, 162*, 1243–1248.
Harding, R. (2017). How Warren Buffet broke American capitalism. *Financial Times*, September 12, 2017.
Harrison, F. (1983). *The power in the land*. London: Shepheard-Walwyn.

Harrison, F. (2005). *Boom-Bust*. London: Shepheard-Walwyn Ltd..
Humphrey, T. (1989). Lender of last resort: The concept in history. *The Economic Review, 75*(2), 8–16.
Jackson, T., & Marks, N. (1994). *Measuring sustainable economic welfare: a pilot index: 1950-1990*. Stockholm: Stockholm Environmental Institute.
Jacobs, J. (1992). *Systems of survival*. New York: Random House.
Jensen, M. C., & Meckling, W. H. (1976). Theory of the firm: Managerial behavior, agency costs and ownership structures. *Journal of Financial Economics, 3*(4), 305.
Jevens, W. S. (1884). *Investigations in currency and finance*.
Juglar, C. (1862). *Des Crises commerciales et leur retour periodique en France, en Angleterre et en les Etats Unies*. Paris: Guillaumin.
Kant, I. (1781). *Critique of pure reason*.
Kennedy, P. (1989). *The rise and fall of the great powers*. New York: Vintage Books.
Keynes, J. M. (1933, July). On national self-sufficiency. *New Statesman and Nation*, 66.
Kim, W. C. & Mauborgne R. (2004). *Blue Ocean strategy*. Harvard Business Review Press, pp. 1–9.
Kirk, M. (2009). The warning. In *PBS public affairs programs*. USA: WGBH/PBS.
Kleinknecht, A. (1987). *Innovation patterns in crisis and prosperity: Schumpeter's long cycle reconsidered*. London: MacMillan Company.
Knight, F. H., & Merrian, T. W. (Eds.). (1945). *Liberalism and christianity*. New York, London: Harper & Brothers.
Kolko, G. (1963). *The triumph of conservatism*. New York: MacMillan.
Kondratieff, N. D. (1926). Die langen Wellen der Konjunktur. *Archiv fur Sozialwissenschaft und Sozialpolitik, 56*, 573.
Kuemmel, R., Ayres, R. U., & Lindenberger, D. (2010). Thermodynamic laws, economic methods and the productive power of energy. *Journal of Non-Equilibrium Thermodynamics, 35*(2), 145–181. https://doi.org/10.1515/jnetdy.2010.009.
Kuemmel, R., Lindenberger, D., & Eichhorn, W. (2000). The productive power of energy and economic evolution. *Indian Journal of Applied Economics, 8*, 231–262. Special Issue on Macro and Micro Economics.
Kuznets, S. (1930). *Secular movements in production and prices: their nature and bearing on cyclical fluctuations*. Boston, MA: Houghton Mifflin.
Lazonick, W. (2014, September). Profits without prosperity. *Harvard Business Review, 92*(9), 46–55.
Lewis, M. (2008). *Panic: The story of modern financial insanity*. NY: W.W. Norton, Penguin.
Littlefield, H. (1964). The Wizard of Oz: A parable of populism. *American Quarterly, 16*(1), 47–58.
Locke, J. (1960). *Two treatises on government*. Cambridge, UK: Cambridge University Press. Original edition, 1698.
Lowrey, A. (2018). *Give people money: How a universal basic income would end*. New York: Crown Publishers.
Mackay, C. (1841). *Extraordinary popular delusions and the madness of crowds, Wiley Investment Classics*. New York: Wiley. Original edition, 1841. Reprint, 1996.

Magdoff, H. (2000). *The Age of imerialism: The economics of US Foreign Policy*. New York: Monthly Review Press.

Malthus, T. R. (1798). An essay on the principle of population as it affects the future improvement of society. In L. D. Abbott (Ed.), *Masterworks of Economics: Digest of Ten Great Classics* (pp. 191–270). New York: Doubleday and Company, Inc.. Original edition, 1798, reprint 1946.

Martin, F. (2014). *Money: The unauthorized biography*. London: Vintage Books.

Marx, K. (1867). *Das Kapital*. German ed. 2 vols.

Marx, K. (1867a). *Das Kapital*. Hamburg: Otto Meissner. German ed. 2 vols.

Marx, K. (1867b). Capital. In L. D. Abbott (Ed.), *Masterworks of economics: Digest of ten great classics*. New York: Doubleday and Company, Inc.. Original edition, 1867, reprint 1946.

Maynard-Smith, J., & Price, G. R. (1973). The Logic of Animal Conflict. *Nature, 246*(5427), 15–18.

Mensch, G. (1979). *Stalemate in technology: Innovations overcome the depression*. Cambridge, MA: Ballinger.

Mill, J. S. (1848a). *Principles of political economy with some of their applications to social philosophy*. London: C.C. Little and J. Brown.

Mill, J. S. (1848b). Principles of political economy. In L. D. Abbott (Ed.), *Masterworks of economics: Digest of ten great classics* (pp. 379–452). New York: Doubleday and Company, Inc.. Original edition, 1848, reprint 1946.

Mill, J. S. (1869). *On liberty* (4th ed.). London: J. W. Parker.

Milner, E. M., & Milner, D. (1919). A scheme for a state bonus: A rational method fr solving the social problem. *The Economic Journal, 29*(114), 241.

More, Sir Thomas. (1516). *Utopia*. Amsterdam.

Morgan, D. & Narron, J. (2015). Crisis chronicles: The panic of 1825 and the most fantastic swindle of all time. *Liberty Street Economics*.

Morton, F. (1962). *The Rothschilds: A family portrait*. New York: Atheneum Publishers.

Mullins, E. (1983). *The secrets of the federal reserve*. Staunton, VA: Bankers Research Institute.

Mullainathan, S., & Safir, E. (2013). *Wgy having so little means so much*. New York: Henry Holt Co..

Mumford, L. (1961). *The city in history: Its origins, its transformations and its prospects*. New York: Harcourt.

Mun, Thomas. (1645 [1946]). England's treasure by foreign trade. In Leonard Dalton Abbott (Ed.), *Masterworks of economics: Digest of ten great classics*, pp. 11–38. New York: Doubleday and Company, Inc.. Original edition, c1645.

Mun, T. (1664). *England's treasure by forraign trade is the rule of our treasure*. London: John Mun.

Narron, J. & Skeie, D. (2014). Crisis chronicles: The credit and commercial crisis of 1772. *Liberty Street Economics*.

Nash, J. F. (1951). Non-cooperative games. *Annals of Mathematics, 54*, 286–295.

Nash, J. F. (1953). Two person cooperative games. *Econometrica, 21*, 128–140.

Nash, J. F. (1950). Equilibrium points in n-person games. *Proceedings of the National Academy of Science, 36*(1), 48–49.

Owen, R. (1813). A new view of society. In L. D. Abbott (Ed.), *Masterworks of economics: Digest of ten great classics* (pp. 343–378). New York: Doubleday and Company, Inc.. Original edition, 1813, reprint 1946.

Paine, T. (1791). *The rights of man*. London: J.S. Jordan.

Piketty, T. (2014). *Capital in the twenty-first century*. Cambridge, MA: Belknap Press.

Quesnay, F. (1766). *Tableau Economique*.

Rand, A. (1964). *The virtue of selfishness*. New York: The New American Library, Signet. New American Library edition.

Rand, A. (1967). *Capitalism: The unknown ideal*. New York: New American Library.

Reinhart, C. M., & Rogoff, K. S. (2009). *This time is different; Eight centuries of financial folly*. Princeton, NJ: Princeton University Press.

Ricardo, D. (1817). Principles of political economy and taxation. In L. D. Abbott (Ed.), *Masterworks of economics: digest of ten great classics* (pp. 271–342). New York: Doubleday and Company, Inc.. Original edition, 1817, reprint 1946.

Ritschl, A. (2012). Reparations, deficits and debt default: The great depression in Germany. In N. Crafts & P. Fearon (Eds.), *The great depression of the 1930s: Lessons for today* (pp. 110–139). Oxford: Oxford University Press.

Rostow, W. W. (1975). Kondratieff, Schumpeter, Kuznets: Trend periods revisited. *Journal of Economic History, 35*, 719–753.

Rostow, W. W. (1978). *The world economy: History and prospect*. Austin, TX: University of Texas Press.

Say, J. B. (1803). *A treatise on political economy*. Philadelphia: Claxton, Remsen & Haffelfinger. Translated by Prinsep. 4th (1821) ed. Paris. Original edition, 1803.

Schumpeter, J. A. (1939). *Business cycles: A theoretical, historical and statistical analysis of the capitalist process*. New York: McGraw-Hill. 2 vols.

Seavoy, R. (2013). *An Economic History of the United States: from 1607 to the present*. New York: Routledge.

Sedlacek, T. (2011). *Economics of good and evil*. New york: Oxford University Press.

Sekera, J. (2016). *The public economy in crisis: A call for a new public economics*.

Shiller, R. J. (2006). *Irrational exuberance* (2nd ed.). New York: Crown Business.

Smith, A. (1776). An inquiry into the nature and causes of the wealth of nations. In M. Lewis (Ed.), *The real price of everything* (pp. 22–652). New York, London: Sterling. Original edition, 1776, reprint 2007.

Standing, G. (2017). *Basic income: A guide for the open-minded*. New Haven: Yale University Press.

Stern, A., & Kravitz, L. (2016). *Raising the floor: How a universal basic income can renew our economy and rebuild the American Dream*. New York: Public Afffairs.

Stigler, G. J. (1987). Frank Hyneman Knight. In J. Eatwell (Ed.), *The New Palgrave: A dictionary of economics*. New York: Stockton Press.

Stiglitz, J. (2015). *The Great Divide: Unequal societies and what we can do about them*. New York: WW Norton.

Stiglitz, J. (2019). *People, power ansd profits: progressve capitalism in an age of discontent*. New York: WW Norton.

Stone, I. F. (2015). *The hidden history of the Korean War, 1959–1951*. Vol. 10, Forbidden Bookshelf, Amazon.Com.

Stout, L. (2012). *The shareholder value myth: How putting shareholders first harms investors, corporations, and the public*. San Francisco: Barrett-KoehlerPublishers, Inc..

Streeter, W. J. (1984). *The silver mania*. Dordrecht: D. Reidel (Kluwer).

Taleb, N. (2007). *The Black Swan: The impact of the highly improbable*. New York: Random House.

Tarascio, V. J. (1988). Kondratieff's theory of long cycles. *Atlantic Economic Journal, 16*, 1–10.

Tawney, R. H. (1926). *Religion and the rise of capitalism*. New York: Harcourt Brace.

Tawney, R. H. (1931 [1952]). *Equality*. 4th ed. New York: Capricorn Books. Original edition, 1931.

Taylor, F. W. (1911). *Principles of scientific management*. New York: Harper & Brothers.

Thierry, F. (2001). La Fiduciarite Ideale: a l'epreuve des couts de production: Quelques elements sur la contradiction fondamentale de la monnaie en Chine. *Revue Numismatique, 157*, 131–152.

Thorp, E. O., & Kassouf, S. (1967). *Beat the market: A scientific stock market system* (1st ed.). New York: Random House.

Triana, P. (2012). *Lecturing birds on flying*. New York: John Wiley.

Turgot, A. R. J. (1776). Reflections on the formation and distribution of wealth. In L. D. Abbott (Ed.), *Masterworks of economics: Digest of ten great classics* (pp. 39–62). New York: Doubleday and Company, Inc.. Original edition, 1765, reprint 1946.

van Duijn, Jaap (2007). *De Groei Voorbij. Over de economische toekomst van Nederland na de booming nineties*: de Bezige Bij.

van Gelderen, J. (1913). Springvloed: Beschouwingen over industriele ontwikkeling en prijsbeweging, de ni. *de Nieuwe Tijd 18*(4,5,6).

van Zanden, J. L. (2009). The long road to the industrial revolution: The European economy in global perspective, 1000-1800. In M. Prak & J. L. van Zanden (Eds.), *Vol. 1: Global economic history series* (1st ed.). Leiden, Boston: Brill.

von Goethe, J. W. (1833/1984). *Faust parts I and II*. Princeton, N.J.: Princeton University Press. (Translated by S. Atkins).

van Parijs, P., & Vanderborght, Y. (2017). *Basic income*. Cambridge, MA: Harvard University Press.

Weatherford, J. (1997). *The history of money*. New York: Three Rivers Press. Paperback edition.

Weber, M. (1902). *The protestant ethic and the spirit of capitalism and other writings*. London: Penguin Books. Translated by Peter Baehr and Gordon C. Wells.

Wilson, D. S., & Wilson, E. O. (2007). Rethinking the theopretical foundation of sociobiology. *Quarterly Review of Biology, 82*, 327–348.

Wilson, E. O. (1975). *Sociobiology: The new synthesis*. Cambridge, MA: Harvard University Press.

Wilson, G. C. (1966). *Adaptation and natural selection: A critique of some current evolutionary thought*. Princeton, NJ: Princeton University Press.

Index

A
Adjustable rate mortgages (ARMs), 148–149, 197
Age of enlightenment
  ancient religious conflicts, 39
  cotton gin, 37
  definition, 27
  "field of the cloth of gold", 35
  history of wars, 39
  humanism, 28
  hybrid government, 29
  industrial revolution, 30
  mercantile policy, 35
  piracy, 35
  popular sovereignty, 28
  privateering, 35
  "scientific method", 27
  Spanish empire, 34
  "Wolfenbüttler manuscript", 30, 31, 33
American Labor Movement, 170
American-style capitalism, 203
"Apex" theory, 136

B
Balance sheets, 57
Bank Charter Act, 126
Bank for International Settlements (BIS), 70, 148
Banking Act law, 102
Bank money, 60
Basel rules, 193
"Before capitalism"
  agriculture, 8
  class system, 15–16
  democracy, 10
  economic theory, 13
  Epic of Gilgamesh, 8
  ethics and morality, 12–14
  fertile crescent, 7
  Holy Land, 10
  LBOs, 9
  modern finance theory, 13
  present-day corporate raiders, 9
  proto money, 10
  social structure and land
    British common law, 18
    chain of command, 18
    citizenship, 18
    Copernican theory, 19
    cultural differences, 20
    feudal system, 17
    Gutenberg Bibles, 20
    human inhumanity, 16

"Before capitalism" (*cont.*)
  knowledge-based power centers, 19
  social divide, 18
  social organization, 16
  trade and manufacturing, 17
 urban civilization, 8
Bill-broking
 Bank Charter Act, 126
 commercial paper, 126
 gold reserve, 129
 LLR (*see* Lender of last resort (LLR))
 LOCs, 125
 nineteenth century Britain, 132
 Overend-Gurney—share prices, 127
 panic of 1857, 127, 128
 panic of 1877, 130
 panic of 1884, 130
 panic of 1893, 130, 131
 panic of 1907, 136
 qianzhuang, 135
 risky investments, 127
 tight money policy, 126
 US financial panic, 130
 Xinhai revolution, 133–135
Bill of Rights, 1
Bitcoin bubble, 158
Black-Scholes-Merton (BSM), 153
Blue Ocean Strategy, 107
Brownian motion, 155
Bubble Act, 116, 126
Bubble-and-bust phenomenon, 114
Bubbles and panics from 1920
 1929 stock market crash, 141
 1997 Asian financial crisis, 154
 1998 Russian financial crisis, 154
 arbitrageur strategy, 152
 avalanche analogy, 158
 benefits of HFT, 156
 Black Monday, 142, 143
 BSM model, 153
 CDSs, 149
 Dot.com bubble, 144–146
 Dot-com crisis, 147
 economic theory, 152
 EMH, 155
 Enron case, 146
 execution risk, 156
 five golden years 2003–2007, 157
 great silver bubble, 143
 hot potato effect, 157
 insurance strategies, 156
 internet, 145
 investment trusts, 139, 140
 Japanese land-stock bubble, 144
 leverage, 141
 LTCM, 153, 154
 mortgage-based bonds, 148
 negative equity, 150
 Nikkei stock index, 144
 price of gold, 142
 silver market, 141
 stock market crash of 2008–09, 152
 stock-price bubble, 144
 Tokyo real estate, 144
 Wall Street financial losses, 149

C
Calvinism, 22
Capital losses, 113
Capitalism
 American-style, 203
 balance of power, 201
 Cold War, 203
 corporate profits, 201
 employee compensation, 201
 financial engineering, 204
 "free markets", 199
 gambling, 205
 inequality, 200, 203
 market-watchers, 204
 non-profit public education system, 201
 traditional economic sense, 200
 traditional finance industry, 205
 US health system, 202

Chinese paper money, *see* Paper money
Coase theorem, 213
Coinage Act, 130
Collateralized debt obligations (CDOs), 194
Commercial paper, 126
Comparative advantage, 46
Competitive Local Exchange Companies (CLECs), 146
Conflict of ideas
 closed circular economy, 87
 competitiveness, 86
 economic theory, 87
 entrepreneur, 88
 factional disputes, 85
 Familistere, 99
 French Revolution, 98
 hub and spoke model, 86
 input–output scheme, 88
 issue of inequality, 91
 labor theory of value, 93
 Marx's political theories, 93
 military force, 91
 "New Moral World", 96
 personalities, 90
 political philosophy, 92
 pre-revolutionary France, 88
 property rights, 88
 social programs, 90
 socialism, 85, 86, 90, 93
 Treaty of Brest-Litovsk, 94
 UBI, 97
 utopian, 95
Copernican theory, 19
Corporate capitalism
 Buffet's first rule, 107
 consumer cooperatives, 102
 corporate profits *vs.* employee compensation, 109
 corporate restructuring, 110
 egalitarianism, 111
 financial tycoons, 104
 fluctuations, 111
 indefinite lifetime, 101, 103
 investment banks, 102
 legal partnership, 101, 102
 modern monopolies/oligopolies, 107
 moment of truth, 105
 organizational forms, 101
 political capitalism, 108
 raiders and activist investors, 109
 returns to capital, 110
 returns to labor, 110
 shareholders, 103
 socialism, 108
 takeover, 106
 trickle down, 108
 type of company, 103
 undervalued, 106
Cost share theorem, 220n3
Credit and banking
 balance sheets, 57
 Civil War, 61
 double-entry bookkeeping, 57
 English-Dutch alliance, 61
 financial market, 59
 Gresham's Law, 59
 LOCs (*see* Letters of credit (LOCs))
 mercantile policy, 63
 national currencies, 58
 silver coins, 62, 63
 stabilization fund, 59
 super companies, 56
Credit default swaps (CDS), 76, 149, 180, 197
Crimean War, 129
Crypto-currencies, 113
Currency Act, 63

Dot-com bubble, 158
Dow Jones Industrial Average (DJIA), 156
Dow Theory, 152

E

Economic cycles
  boom and bust cyclicity, 161
  cyclical theory, 163
  economic theory, 162
  fixed investment cycle, 162
  Goodwin and Minsky cycles, 162
  Goodwin bubble mechanism, 166
  infrastructural investment
      cycle, 162
  inventory cycle, 162
  investment-debt cycle, 166
  "Long Economic Cycles", 164
  Juglar (business) cycle, 162–164
  Kondratieff cycles, 164, 165
  Kondratieff waves, 164
  long-wave technological cycle, 162
  "Measuring Business Cycles", 163
  minor cycles, 162
  "Principles of Political
      Economy", 162
  Schumpeterian wave, 163
  technical change and innovation/
      diffusion, 166
Economic theory, 87
Efficient Market Hypothesis
    (EMH), 155
Elliot Wave Theory, 152
Evolutionary game theory
    (EGT), 80

F

Fear Index, 158n5
Federal Reserve Bank (FRB), 110
"Field of the cloth of gold", 35
Financial crises, 114
Florida land bubble, 120
Florida real estate bubbles, 119
Fourth Coinage Act, 130
"Free rider" issue, 213
"Free trade" agreements, 170–172
French Revolution, 98, 199

G

General Agreement on Tariffs and
    Trade (GATT), 171
Glass–Steagall banking law, 206
Glass–Steagall law, 75
Gold
  coins, 47, 48, 51, 69
  currency, 71
  Dawes Plan, 69, 70
  hyper-inflation, 67
  quantitative easing, 70
  store of value theory, 71
  Treaty of Versailles, 66, 69
  World War, 65
Gold-bugs, 65
Gold bullion standard, 69
Greenmail, 178–179
Gresham's Law, 59
Gross national product (GDP), 110

H

High frequency trading (HFT),
    155, 156
Hybrid government, 29

I

Independent Local Exchange
    Companies (ILECs), 146
Industrial ecology, 4
Industrial metabolism, 4
Industrial revolution, 30
Industrial Workers of the World
    (IWW), 86
Initial public offering (IPO), 186, 196
Intelligent design, 3
Investment trusts, 139, 140, 176, 177

J

Joint Stock Company Act, 103
Junk bonds, 179

Index

L

Labor movement
  employers and employees
      balance, 169
  "free trade" agreements, 170–172
  globalization, 172
  LBO movement and
      movement, 169
  strike-breakers, 170
  US employee compensation, 170
  Walmart business model, 174
  Walton's successful strategy, 173
  Washington Consensus, 173
LBOs movement, 169
Lender of last resort (LLR), 119, 127, 128, 132, 144, 149
Letters of credit (LOCs), 55, 56, 60, 125
Leveraged buyout (LBO)
  ARMs, 197
  Basel rules, 193
  buyouts, 192
  "golden parachutes" gimmicks, 193
  ideas and intellectual property, 198
  IPO, 193
  junk bonds, 191, 192
  mortgage-based bonds, 194
  mortgage default rates, 196
  negative-sum (lose-–lose) game, 194
  parasitic financial sector, 196
  "poison pills" gimmicks, 193
  private equity, 192
  securitization gimmick, 196
  South Sea Bubble, 194
  stock-market-based incentives, 193
  stock-market losses, 198
  SVM theory, 198
  TARP, 197
  western capitalism, 191
  "zero zero-sum" game, 194, 195
Leveraged buyouts (LBOs), 9, 179
Limited Liability Act, 103n2, 104
Long-Term Capital Management (LTCM), 153, 154, 179

M

Mahayana Buddhism, 13
Margin loans, 176, 177
Marxism, 84
Marx's Theory of Crisis, 161
Mercantile policy, 35, 63
Military-industrial complex, 42
Mint Act, 130
Mississippi Bubble, 115
Money
  bronze coins, 47
  bullion, 51
  coinage, 47
  copper coins, 47, 49
  electrum, 47
  European passion, 51
  geographical and topographical differences, 45
  gold (*see* Gold)
  Hall–Heroult process, 52
  intrinsic value, 65
  Joachimsthalers, 49
  medium of exchange, 46
  natural resources, 45, 46
  printed paper (*see* Paper money)
  silver coins, 51
  taylorism, 53
Mortgage-based securities, 180
Mosaic law, 12

N

Navigation and Trade Acts, 64
Negative income tax, 208, 211
Negative-sum (lose-–lose) game, 194
Newton's theory of gravitation, 2

P

Paper money, 73–76
Peak price-earnings (PE), 147
Ponzi scheme, 116
Portfolio insurance, 156

"Principal-agent" theory, 181, 185
Privatizing profits and socializing losses
  arbitrage, 179, 180
  bank bailouts, 180–181
  greenmail (*see* Greenmail)
  investment trusts, 176, 177
  junk bonds, 179
  LBOs, 179
  margin loans, 175–177
  share buybacks *vs.* R&D, 186–189
  SVM (*see* Shareholder value maximization (SVM))
  unequal voting rights, 176
Protestant Ethic, 24, 211
Protestant Reformation
  Calvinism, 22
  commercial syndrome, 23
  European rivals, 25
  guardian syndrome, 23
  Indian Mutiny, 25
  industrial revolution, 25
  society and government, 24
  spirit of capitalism, 24

Q

Quantitative easing (QE), 70, 207
Quesnay's Tableau Economique, 89

R

Railway share price bubble, 121–123
Republican Tax Cuts and Jobs Act, 187
Resources for the Future (RFF), 4
Reward Work Act, 189
Russian Revolution, 18

S

Sarbanes-Oxley Act, 147
Scarce resource taxes, 220
Shareholder value maximization (SVM), 181–186
Sherman Anti-Trust Act, 107
Sherman Silver Purchase Act, 130, 131
Social reformers, 2
Socialism *vs.* capitalism
  economic growth, 83
  intellectual revolution, 78
  land ownership, 78
  Lebensraum, 77
  modern governments, 83
  prospect theory, 80
  re-investment of profits, 81
  transfer of wealth, 82
  utopian socialism, 79
South American investment bubble, 118, 119
South Sea bubble, 116
Store of value theory, 71
SVM movement, 169

T

Taft-Hartley Act, 206
Taxpayer Relief Act, 144
Taylorism, 53
Temple of Solomon (Templars), 55
Tight money policy, 126
Townshend Acts, 64
Troubled Asset Relief Program (TARP), 149, 197

U

Universal basic income (UBI), 97, 98
  1980, 209
  2016, 209
  anti-war movement, 208, 210
  borrow and/or print money, 207
  carbon (exergy) taxes, 219, 220
  costs and benefits, 221
  "Dot Com" stock market boom, 218
  earned and unearned income for tax, 217
  economic impact, 221

educational and health-care programs, 210
excise taxes, 220
generation of "Georgists", 216
global nuclear war, 212–216
government services, 216
human right, 212
hydrocarbons and embodied carbon product, 219
military expenditure, 217
National Security, 212
negative income tax (*see* Negative income tax)
negative reinforcement, 217
nineteenth century, 209
people decisions, 211
social dividend, 219
social justice, 211
trade war, 208
unemployment issue, 210
VAT, 221
wealthy people, 210
US Telecommunications Act, 145

V

Value added tax (VAT), 221
Value at Risk (VaR), 156

W

Wealth, 41–43

W

"Zero zero-sum" game, 194, 195

GPSR Compliance
The European Union's (EU) General Product Safety Regulation (GPSR) is a set of rules that requires consumer products to be safe and our obligations to ensure this.

If you have any concerns about our products, you can contact us on

ProductSafety@springernature.com

In case Publisher is established outside the EU, the EU authorized representative is:

Springer Nature Customer Service Center GmbH
Europaplatz 3
69115 Heidelberg, Germany